D0340913

SYLVAN OAKS LIBRARY
6700 AUBURN BOULEVARD
CITRUS HEIGHTS, CA 95621
APR   1999

THE HUNTING APES

The Hunting Apes

*Meat Eating and the
Origins of
Human Behavior*

CRAIG B. STANFORD

PRINCETON UNIVERSITY PRESS
PRINCETON, NEW JERSEY

Copyright © 1999 by Princeton University Press
Published by Princeton University Press,
41 William Street, Princeton, New Jersey 03540
In the United Kingdom: Princeton University Press,
Chichester, West Sussex

All Rights Reserved

Library of Congress Cataloging-in-Publication Data
Stanford, Craig B. (Craig Britton), 1956–
The hunting apes : meat eating and the origins
of human behavior / Craig B. Stanford.
p.   cm.
Includes bibliographical references and index.
ISBN 0-691-01160-5 (cloth : alk. paper)
1. Apes—Behavior. 2. Apes—Food. 3. Human
evolution. 4. Hunting and gathering societies. I. Title.
QL737.P96S73   1999
599.88′153—dc21   98-34872

This book has been composed in Palatino
Designed by Jan Lilly

The paper used in this publication meets the
minimum requirements of
ANSI/NISO Z39.48-1992 (R1997)
(*Permanence of Paper*)

http://pup.princeton.edu

Printed in the United States of America

10   9   8   7   6   5   4   3   2   1

For Adam

Contents

Preface and
Acknowledgments

It seems to me that of all the questions that await answers in science, two are most central. The first deals with the nature and origin of the universe and is the realm of astronomers. The second is who we are and where we come from, and that is what this book is about. I began to think about writing this book while I was studying chimpanzees and their prey in East Africa in the early 1990s. Nearly every journal article on human origins that I had ever read paid homage to the importance of chimpanzee behavior in understanding the lives of early hominids. But these papers were just lip service; the authors rarely took very seriously the integration of the different routes to understanding human origins. I felt much would be gained by bringing together the information about our fossil ancestors, modern primates, and modern human-hunting-and-gathering societies. In 1995, while working on another book about chimpanzee predatory behavior, I was invited by Princeton University Press to write precisely that sort of book, to become part of the Princeton Trade Science series. For this invitation I thank PUP, particularly science editor Jack Repcheck.

The reader needs no background in human evolution or primate behavior to appreciate the significance of what I will describe about who we are and where we've come from. The figures and tables are few in number and serve to illustrate important points in the text. A portion of *The Hunting Apes* stems from my own research on chimpanzees and other primates. However, most of the information contained in the book is secondhand; the reader may consult the notes at the back of the book for the original sources.

Many people have made suggestions for revising or improving the book's content. I am grateful for comments, criticism, and helpful discussions on individual chapters from John Allen, Christopher Boehm, Richard Byrne, Elan Glasser, Kristen Hawkes, Adriana Hernandez, Alison Jolly, Richard Malenky, Alexander Moore, Erin Moore, Jim Moore, Martin Muller, Thomas Plummer, and Nancy Thompson-Handler. My thanks especially to Alison Jolly and Jim Moore, who read and critiqued the entire manuscript, as well as to Jack Repcheck.

For permission to study chimpanzees in the early and mid-1990s in Gombe National Park in Tanzania, I thank the Tanzanian Commission for Science and Technology, Tanzania National Parks, and the Serengeti Wildlife Research Institute. For permission to study chimpanzees and mountain gorillas in Bwindi-Impenetrable National Park in Uganda, I thank the Ugandan National Research Council, the Ugandan Wildlife Authority, and the Institute for Tropical Forest Conservation. I am especially grateful for assistance and companionship over the years in these two projects to Anthony Collins, Caleb Gambaneza, Michele Goldsmith, Jane Goodall, Maryke Gray, Simon Jennings, Msafiri Katoto, Richard Malenky, Hilali Matama, Eslom Mpongo, Keith Musana, Nkurunungi John Bosco, William Rutaro, Nancy Thompson-Handler, and Janette Wallis. I also thank the Jane Goodall Institute and the Jane Goodall Research Center at the University of Southern California for access to videotaped records of hunting behavior and other aspects of chimpanzee life. Finally, this book and the research that contributed to it could not have been done without the long-term support and understanding of my wife, Erin Moore, and of our children Gaelen, Marika, and Adam, and I thank them.

South Pasadena, California
June 1998

THE HUNTING APES

The Indelible Stamp

Man still bears in his bodily frame the
indelible stamp of his lowly origin.

Charles Darwin,
The Descent of Man (1871)

The hunters and I are up early. Nine of us—eight of
them plus me as observer—are in a small forested valley
among rugged hills in a remote part of East Africa. We
left together in the darkness of early morning, and now
as daylight comes the band of hunters stops on a grassy
hillside overlooking a lake. They gather a breakfast of
berries and leaves while I sit nearby eating a granola
bar. We do not speak, nor can we speak any language the
other would understand. I am simply following quietly
and taking notes. During breakfast the hunters hear
calls from their neighbors to the north, and set off to
meet them. We cross a series of ridges, and as the group
traverses a stream bed and climbs the valley slope on the
other side we see and hear a group of monkeys feeding in
a stand of small trees. The monkeys are social and noisy,
clamoring about and leaping around in the lower
branches. The hunters quickly assess the situation and

run to the base of the trees; several begin to climb up toward the monkeys while others remain on the ground below, scanning the treetops. A large monkey falls from the tree while trying to escape and thuds into the dry leaves at my feet. A hunter rushes to grab it, then thumps it against the ground until it is dead. A moment later he steals another hunter's kill with impunity and stands in front of me gripping each monkey in a fist. After several minutes of frenzied action the hunt ends with five monkeys caught. Everyone then sits down around the base of the tree, feasting on the meat they have caught. The hunters politick throughout the meal, sharing and swapping scraps of the much-desired meat. For more than two hours they eat the monkeys, and the noise of bones crunching and contented grunting is all around me. The hunters share the bounty with one another, finish off most of the meal, and then nap for an hour. Every bit of the carcass—including bones and skin—is eaten raw. After they are sated and rested, they get up as if on cue and walk off in search of more food.

This event took place not among a group of African hunter-gatherer people, such as the Hadza of northern Tanzania or the !Kung of the Kalahari desert, but among wild chimpanzees. In forests of western Tanzania and across equatorial Africa, these apes include the meat of other mammals as a small but regular part of their diet. The parallels to what traditional human societies do with meat are often striking. Traditional foraging people ("hunter-gatherers") subsist on the natural products of the landscape and get by with minimal technology, gathering plant foods and hunting for small animals.

In this book I argue that the origins of human intelligence are linked to the acquisition of meat, especially through the cognitive capacities necessary for the strategic sharing of meat with fellow group members. Important aspects of the behavior of some higher primates—hunting and meat sharing and the social and cognitive skills that enable these behaviors—are shared evolved traits with humans and point to the origins of human intelligence. This does not mean that there is an instinctive desire to hunt on the part of all modern humans; only a small percentage of people in industrialized countries have ever hunted for anything that's alive. Instead, the intellect required to be a clever, strategic, and mindful sharer of meat is the essential recipe that led to the expansion of the human brain.

Chimpanzees hunt and eat the meat of a variety of mammals. They are skilled makers and users of tools. These apes and their closest relatives have large brains and an intellect that surpasses that of all other nonhuman animals. They are funhouse mirrors of our ancestry; the same stock produced us, but with a filter of millions of years of adaptations that occurred during the history of each lineage. Chimpanzees, along with the other great apes—the bonobo, gorilla, and orangutan—illustrate how evolution can mold a highly intelligent animal that lives in a complex forest environment and an even more complex society. There is only one animal of greater intelligence, and it also lives in an incredibly intricate web of social relationships, navigating its way through life using group-mates as support systems and as tools to be manipulated. This other animal, of course, is humankind.

A second key piece of evidence about the behavior patterns that made us human is that our ancestors foraged for meat. The fossil record contains evidence of increasingly sophisticated tool manufacturing beginning some two and a half million years ago, just as the human brain began to approach the size threshold that is considered human. Reseachers believe that this tool use facilitated an increase in the importance of meat in the early human diet. Exactly when did meat become an important part of the diet, and how was it obtained? Were early humans savage and cunning hunters, or clever but weak scavengers? How important was meat in the diet as our ancestors' lineages evolved and diversified, and how could the eating and sharing of animal prey have contributed to the expansion and reorganization of the human brain and cognition? What are the nutritional and social roles of meat in traditional human societies? These are questions to which anthropologists studying the fossil record have few answers.

I also examine traditional human societies using the same Darwinian paradigm that has provided answers to key questions about animal behavior. Comparing the behavioral ecology of humans living in very traditional settings to nonhuman animal ecology is an inquiry into whether both are driven by the same principles of natural selection. Because the direct evidence of early humanity in the fossil record is and always will be scanty—full of bones but lacking in flesh, both literally and figuratively— information on living human and nonhuman meat eaters is very important. There is much to be learned from modern hunting people in this regard. Modern foraging people are not relicts of the past. They have lives and societies with as much cultural sophistry as any other group of modern humans. But technologically they tend to be simpler, allowing us to see how people who need to subsist from their forest or savannah worlds can do so. This interaction of ecology and behavior provides a backdrop against which the potential range of ecological adaptations of ancient humans can be considered.

If not for the anthropocentrism of the earliest taxonomists—the scientists who devised the naming system we still use to classify living things— humans and apes would be grouped together because of our many shared traits. We three hunting apes—chimpanzees, ancestral hominids, and modern foraging people such as the !Kung or Aché—provide a frame of reference of our evolutionary history and therefore the roots of human behavior. I am not the first anthropologist to address these issues, although this is the first account to integrate modern evidence from my three areas of interest. The role of meat in the lives of early hominids has been viewed at times as crucial, at other times as minor, and at still other times as nonexistent in different eras of anthropological thought. With fossils and human foragers providing supporting evidence for what we know about great apes, we can consider a detailed triptych of hunting, scavenging, and meat sharing, all aimed at exploring the origins of human behavior.

Man the Hunter Revisited

In 1990 I wrote a short paper with a colleague lamenting the way in which human evolutionary scholars were getting away with erecting male-centered portraits of our early human ancestors, hiding behind the mask of Darwinian principles.[1] Shortly thereafter I went to Tanzania and watched wild chimpanzees for the first time. What I saw was a shock to my values regarding a gender-balanced ethics of behavior. Male chimpanzees brutalize females routinely, coercing them for sexual access and punishing them when they don't receive a desired mating. Some females certainly wield power as the matriarchs of powerful family lineages, but even high-ranking females submit to the lowest-ranking males much of the time. I came to see this ape society as one in which might makes right, a pervasive pattern in human patriarchal societies as well. In both chimpanzee and human societies, the control of meat contributes to a might-makes-right form of patriarchy. This led inevitably to my fascination with the notion of Man the Hunter, or perhaps Man the Meat-possessor. As I sat in an African forest collecting data on chimpanzee hunting and meat-sharing behavior, I did not realize that I was entering an academic jungle considerably wilder than the one my chimpanzees inhabited.

A previous generation of scholars sought to explain the role of hunting in the origins of human behavior and intelligence. It was called "Man the Hunter" after a volume that grew out of a conference on the behavior of modern hunting and gathering peoples. A paper by Sherwood Washburn and Chet Lancaster in that book attributed many aspects of modern human social behavior and intellect directly to a history of hunting large animals. The coordinative and communicative abilities that are fundamental to the success of a cooperative hunt were linked to the increasing role of meat eating. In subsequent years this model of human origins was roundly attacked, in part on factual grounds and in part due to the inherent and un-acknowledged biases it was said to contain. Man the Hunter is infamous in anthropological circles, and the scholarly reactions against it cannot be overstated. Feminist anthropology as a discipline grew in part out of the anger over the gender ineq-uities inherent in Man the Hunter. This is because in its original formulation the model of human brain expansion based on male-driven activities was tinged with sexism and blatantly ignored evi-dence of the important role of women in acquiring protein in most societies.

Recent information on the behavior of human hunters and nonhuman primate hunters and from the fossil record points to the crucial role of meat eating, and especially meat sharing, in the roots of human behavior and intellect. Males typically ob-tain meat in human and nonhuman primate soci-eties and then attempt to use it to manipulate or control females. Socially correct gender politics not-withstanding, this is an empirical and demonstrable reality. While early views of "Man" were deeply flawed by a tendency to ignore the female side of human evolution, and by a concomitant ignorance of data on women in human societies and of female nonhuman primates, the central importance of meat acquisition and meat sharing in modern and ancient human societies is simply undeniable. Meat, not only as a nutritionally desirable food item but also as a social currency that is controlled by males and therefore is a tool for the maintenance of patriarchal systems, plays an essential part in the social systems of both traditional human and some nonhuman primate societies. The main thrust of my book is this: that Man the Hunter was fatally flawed, first by its emphasis on the role of cognition in meat acquisition rather than meat sharing, and second by its unconscious ignorance of the role of females in the meat-control system. Correcting these two errors leads us to establish a new framework in the light of modern evolutionary theory and of current views of the roles of women and men in human societies, past and present.

THE EXPANDING CIRCLE OF DARWINISM

Educated and otherwise enlightened people everywhere are reluctant to accept the full extent of the evolutionary process in the natural world. Once it was widely believed that while physical laws governed the movements of bodies in our solar system, life on earth was attributed to a divine power. Later, in a post-Darwinian world, this split worldview came undone, but many continued to believe that while all animals and plants were the products of natural selection, the human species had somehow been exempt from it. Perhaps this can be attributed to our short mammalian life spans: eighty years is too short a time in which to touch and feel the evolutionary process in the way that we can sense the presence of other shorter-lived physical and biological laws of nature. Still later, many philosophers of the human condition acknowledged the humble organic origins of humanity while continuing to maintain that the origin of human consciousness is the product of supernatural forces. The nature of the human mind is today a subject of great debate. There are those who see the mind as the sum of our long history of incrementally more complex neural inner workings, and those who do not accept that any number or arrangement of neurons could account for the tasks our brain is able to accomplish. The philosopher Daniel Dennett, one of a handful of scholars in the humanities who have turned to Darwinian principles to make sense of the human psyche, refers to the explanatory power of natural selection as a "universal acid." It is a conceptual framework that explains everything in its path, and the last stop is human self-awareness and intelligence. As the most intelligent creatures on the planet besides humans, the apes offer us key insights into the makings of our own intellect.

"The Indelible Stamp" refers to the final sentence of Charles Darwin's second great book, *The Descent of Man* (1871), in which he took his principles of evolution through natural selection and applied them to humans. Our bodies are a mosaic of evolutionary influences that have acted on our ancestors at many stages in our history. For example, the grasping hand was molded in the last days of the dinosaurs some 65 million years ago; our upright posture evolved some 6 million years ago; our enormous brain ballooned to its present size only some 100–200,000 years ago. Together, the package has become what we, for the purpose of biological classification, consider human. But it is not only our anatomy that is the product of evolution. Our sociality, the most basic primate behavioral adaptation, is a product of our status as a higher primate. While we learn to be members of one culture or another through learning, our social nature itself is as basic a primate trait as breathing. A whole range of human social behaviors, from mother-infant bonding to corporate striving to choosing a mate to sexual jealousy, is influenced by evolved tendencies to respond within a certain range of emotional and behavioral reactions to particular situations. Eyebrows are raised when we consider human behavior to be motivated only by our human species centrism.

This Darwinian paradigm is of fundamental importance in tearing down old perspectives about meat eating and its role in human origins and in forging a new synthesis. Without a framework rooted in the realities of what animals, including human animals, do, we are left with stories told through the ages and all their inherent biases and flavorings. By considering what apes do, what modern humans do, and what early hominids probably did, we may come up with an integrated view of human behavior.

Man the Hunter and Other Stories

It is necessary to remember that fossils were alive when they were important.

S. L. Washburn,
in *Anthropology Today* (1967)

Around five or six million years ago, a hairy, one-meter tall creature that looked much like an upright chimpanzee left the security of the woodlands for life on the open savannahs of eastern Africa. Physically defenseless but gifted with a sharp mind, the species had carved a niche for itself by becoming bipedal, allowing the creatures to travel efficiently for long distances over open ground. Its members ate mostly fruits and leaves but also included increasing quantities of meat, both hunted and stolen from animal carcasses that they found on the savannah. Eventually they learned to modify stones into tools, which over time became effective butchering implements, and perhaps also weapons. Intelligence and sociability were their most valuable assets in coping

*with the risks of a dangerous world. They therefore lived
in social groups or perhaps in monogamous pairs for the
safety of numbers. These first hominids flourished, and
their descendents became ever larger-brained and more
human, eventually evolving into modern Homo sapiens.*

The above account of human origins has been the
standard version that students of human origins
as well as the public have learned and recited for
many years, but you should not find it very con-
vincing. Our knowledge of what the earliest members of our hominid family were like is itself still
evolving. The evidence for bipedalism arising in a
fairly treeless savannah niche is, for example, considered tenuous.[1] Did the earliest hominids (members of the human family) hunt, or did they scavenge to obtain their meat? What precise factors
brought about the dramatic expansion in brain size?
For every simplistic portrait of human origins such
as the one above, there are dozens of researchers
eager to tackle other untested assumptions. Some of
these debates over our origins are quite fierce and
have profound implications for understanding who
we are.

The study of human origins is like the plot of the
Japanese film *Rashomon*. A brutal crime has been
committed and observed by a number of eyewit-
nesses. A suspect is apprehended, but each eye-

witness account of the crime in question differs in some critical way. Even though the facts have been established, the police are unable to reconstruct the event to their satisfaction. The crime is clear but the lines of evidence do not converge due to the prism of circumstance and perspective through which the observers' differing accounts pass. This is the situation in which scientists studying human origins find themselves; the bits of evidence from primate behavior, fossils, and other sources do not always build a consensual picture. Instead, they often point to a variety of possible scenarios, all of which probably contain important elements of the truth.

THIS OLD HOUSE

The difficulty in studying human origins lies in our narrow view of the ancient world. We are constrained by our firsthand knowledge only of the present, in which the why and wherefore of the past are lost. Go into an old house and look around. If, like most older homes, it has been remodeled and refurbished many times over its long life, you will be confronted with a jumble of architectural and furnishing styles. The Victorian fixtures in the living room, the Craftsman sideboard in the dining room, and the 1960s kitchen remodel all coexist in one place. Modern heating and wiring systems are built on top of the original, and beneath the wallpaper there are other layers from previous chapters in the house's life. These changes obscure the original appearance quite effectively. Today the house is a mosaic reflecting a long history of styles and decors, *none of which could have been predicted by the previous generation's designers*. A casual observer can no more look back from the 1990s to see the house clearly in its original form than he could have stood in the kitchen in the 1890s and predicted the course the house would take in the century to come.

The lesson of the old house is a good one for those who theorize about the primate origins of our own behavior and anatomy. Each piece of furniture and each component of the floor plan is one part of the jigsaw puzzle of evolutionary function and form that natural selection molds. What's more, the puzzle is internally interactive. That is, when a new piece is added, other pieces have to accommodate and be accommodated. When bipedal posture is adopted, the circulatory system, the spinal column, the diet and foraging behavior, and even the mode of social interaction also change. This makes assembling the puzzle retrospectively an enormous challenge. In constructing theories of our origins, we amass diverse evidence from fossils, modern human behavior, the behavior and anatomy of living primates, and from genetic studies to develop conceptual models of what our earliest ancestors were like. But in the end we often create scenarios that rely heavily on the behavior of just one closely related species, which becomes an analogy for the form that our ancestors may have taken.

The chimpanzee model has long been foremost among these analogous models due to its obvious appeal. We are physically very similar to these apes and studies show our DNA sequences to be about 98.4 percent the same; this makes chimpanzees more closely related to us than they are to gorillas. But the 1.6 percent different DNA is enough to make us distinct, having followed our separate evolutionary paths for 6 million years. We can easily see how different modern humans are from our ancestors. Modern chimpanzees, on the other hand, appear to have remained more similar to our common ancestor. But it would be a mistake to assume that early humans were very similar to chimpanzees simply because they looked more like chimps than like modern people. There have been, in effect, 12 million years of evolution (6 along each prong of the fork) separating humans and chimpanzees—long enough for profound changes to have occurred. The ape and the human were quite similar at the earliest stages of the early hominid-chimpanzee ancestor branching, when the lines had been going their own way for no more than 2–3 million years. We must compare the modern chimpanzee with the likely anatomy and behavior of our early common ances-

tor that lived before the node occurred in our families' branches. It is this gap that we are trying to
span when reconstructing our family roots.

What is a model of human evolution? Primatologist Jim Moore distinguishes a model from a theory.
A model is an explanation that uses one member of
an analogous pair as a point of reference in understanding the bigger picture.[2] Thus craters on the
moon can be used to model the way in which celestial objects crash into celestial bodies in our solar
system. We can study the moon in detail because of
its proximity and then use a lunar model for crater
formation. In time, space probes may prove this
model to be completely wrong if, for instance, we
learn that craters on other moons in the solar system are formed by other means. The lunar model is
an analogic model; it uses one well-known example
to predict conditions elsewhere. A theory, by contrast, should be an intellectual construct that uses
facts from disparate sources to build a prediction
about some natural phenomenon. For instance, one
might theorize that based on studies of the velocity
and impact rates of meteors that have been recorded
throughout the solar system, the appearance of
craters on celestial bodies everywhere should be
similar. This is different from the lunar model of
crater formation because it casts its net more widely
for relevant data to test the hypothesis about craters. Moore points out, however, that, in practice,

models and theories are often treated as one and the same.

A theory ought to have predictive power. In the physical sciences, researchers claim they have proved the existence of objects, whether elementary particles or celestial bodies, that have never been observed visually. Showing theoretically that a black hole or brown dwarf ought to exist is enough to provoke other researchers to take up the search. This is because of the consistent and predictive laws of the physical world. Human evolutionary scientists hypothesize retrospectively what extinct creatures may have been like, choosing from a wide range of possible adaptations. Researchers have attempted to create predictive theories, plausible scenarios, or even simply stories that contain some well-informed guesses about a long-gone reality. But in truth, evolutionary models are, in spite of their tremendous power to explain past history, not predictive. Models of the evolutionary process can describe and explain the history of organic change in a lineage even though they could never have predicted that those changes would occur.

PRIMATE MODELS OF EARLY HUMANS

When we want to build a model of something that no one has ever seen, it helps a great deal to have another structure on hand that is similar to what is being modeled. However, having that similar-but-not-the-same model handy may mislead us in important ways. We could use the principles of primate ecology combined with what we know about the relationship between anatomy and behavior to reconstruct the social lives of early hominids. This approach, however, is fraught with uncertainties. If the solitary-living orangutan were not alive today, primatologists examining ancient orangutan skeletons would be perplexed to note that males were dramatically larger than females, which is typical of polygynous, group-living primates but completely unexpected for solitary ones. Similarly, no primatologist could determine the behavioral differences between the chimpanzee and the bonobo based on the animal's anatomy without having a fully fleshed version to study in life as well. In fact, if the chimpanzee and bonobo existed only as fossils they might be mistaken for one species because of their anatomical similarity. These are important reminders of the limitations in modeling the behavior of extinct animals. When a researcher constructs a grand conceptual model of early human evolution, she or he has no choice but to resort at some point to the use of a strong analogy with some living, well-studied primate relative.[3] There are four great apes, and any one of them could be a good analog for the human ancestor. But of course the human ancestor was not exactly like any of them, and perhaps was so different that to use any as exemplars may be counterproductive.

We can construct a portrait of a hominid ancestor based on the behavior of one or a few living and highly analogous species. Most models that attempt to use empirical evidence from a range of disciplines implicitly use a chimpanzee model as the psychological starting point. Primatologist Richard Wrangham has argued that of the four great apes, the chimpanzee is the best model of our own origins because it has more features that link it with gorillas and humans than with bonobos. He posits the chimpanzee as the direct best analogic model for the last common ancestor, citing molecular, anatomical, and behavioral links between chimpanzees and gorillas that leave bonobos as an outgroup.[4] Because we know that we are more closely related to chimpanzees than to gorillas, the former would then be the last living word on our ancient selves. Adrienne Zihlmann and her colleagues, meanwhile, have long proposed the bonobo as the best model for an early hominid, citing morphological similarities in this ape to the human bipedal adaptation.[5] Other researchers have seen these similarities only as interesting but coincidental parallelisms; the apparent adaptation to bipedal posture may be an adaptation to an arboreal lifestyle.[6] Bonobos are probably not a more appropriate model for the last common ancestor with hominids than chimpanzees.

It is impossible to overstate the role of the chimpanzee and bonobo in the evolution of theories about human origins. Understanding the lives of extinct forms represented only by bones would be vastly more difficult in a world in which paleoanthropologists had access only to the newly discovered hominoid fossil and had never seen living individuals. Live animals provide us with examples of the range of possible adaptations for feeding, ranging, territoriality, mating, offspring rearing, and a variety of other behaviors without which there would be no starting point for reconstructing hominid lifeways. Not surprisingly, most self-proclaimed conceptual models have settled on a very chimpanzee-like creature as the presumed common ancestor of all the hominids.[7] Only by comparing the ecological and anatomical features of the living apes can we hope to distinguish the range of possible adaptations from those that the extinct species was likely to have had. The problem of extrapolating from living to extinct forms is compounded by the absence of fossil African apes that would provide physical evidence of the evolution of the chimpanzee-bonobo lineage after their divergence from the shared ape ancestor.

Perhaps the greatest misconception about the use of a chimpanzee model of human evolution is the often repeated notion that using a chimpanzee analogy limits our perspective of the unique traits that early hominids would have possessed. Andrew Hill speaks for many paleoanthropologists in stating that "pretending that the very early hominids are almost exactly like modern chimpanzees, or any other particular animal, seems to me a dead end. Work is then devoted to confirming this view of the past, and this practice prevents the detection of differences."[8] Hill is mistaken: if anything, using a chimpanzee model keeps us from forming a cardboard view of any extinct hominid species. The earliest primate researchers used to speak of "the monkey" to refer to all primates, in the days before we appreciated the dramatic diversity of mating systems and other behaviors among the primate order. Likewise, William McGrew has chastised us for thinking of "the chimpanzee" by pointing out that the degree of cultural diversity among chimpanzee populations across Africa prevents us from generalizing about the species' tool-use capabilities, hunting styles, and so on. And in the same way, paleoanthropologists err when they speak of "the australopithecine" as though each species of early human was monolithic. In all likelihood, there were populations of *Homo habilis* that hunted avidly for small mammals, and other populations living at the same time 100 kilometers away that were scavengers. This is the nature of the diversity of chimpanzee societies, and there is good reason to think it would have characterized early human populations as well.

Other theorists have used great ape societies as referent points for reconstructing aspects of the earliest hominid's social behavior and ecology. These models have focused on different aspects of species-specific traits as critical features also present in the common ancestor. Richard Wrangham compared the social systems of the African apes to ascertain the behavioral ecology of the common ancestor; he concluded that polygynous, male-bonded kin groups were the core of the common ancestor's society. He found inconclusive evidence about territoriality and male philopatry.[9] Michael Ghiglieri conducted a similar analysis using only the chimpanzee and bonobo and considered male reproductive strategies and intense male kin group-enforced territoriality to be the core features around which a model of the stem hominid should be built.[10] These papers focused on male bonding in relation to male reproductive strategies, because the emerging consensus on the nature of both chimpanzee and bonobo society is that they are composed of male-bonded kin groups. Only a few researchers have considered female *Pan* behavior in such models of early hominids. William McGrew posited the female chimpanzee as a prototype of an early hominid female; he saw active mate solicitation in a polygynous setting as evidence of females that were active reproductive strategists rather than passive receptacles for males' reproductive ambitions.[11]

Models are of course only as well supported as the information on wild apes that is used to build them. In the 1970s, when a debate over whether early humans were hunters and scavengers was brewing, the archaeologist Glynn Isaac built a seminal model of early human behavior based on the sharing of food after butchering the carcasses of savannah ungulates. Isaac rejected the behavior of modern chimpanzees in building this model. He did not know then that meat consumption by chimpanzees is as frequent as we know it is today. Isaac considered chimpanzee hunting to be more similar to what we would call gathering in traditional human societies because it involved only small packages of protein.[12] We know today that the cumulative amount of meat can be quite high because of the tendency for chimpanzees to capture several monkeys in a single hunt. Chimpanzees share meat nepotistically but they do not store food for later consumption; hand to mouth is the rule. Isaac would have been forced to frame the sharing hypothesis differently had he known of the more recent discoveries about chimpanzee behavior.

Analogic models do not necessarily have to be built upon the biology of our very closest relatives. The first modern use of a referent species for early humans was in fact a monkey, the savannah baboon. Irven DeVore and Sherwood Washburn published a series of papers in the early 1960s that focused on the role of the male baboon and its status-seeking, mate-guarding, and sometimes meat-eating nature. These findings held sway for many years in modeling our view of what the most important factors driving primate societies were. Eventually it was pointed out that female baboon behavior is as interesting and important as that of males, but male behavior continued to dominate the research agenda on baboons until recent decades. We saw ourselves as baboon-like, and so our models of our ancestors suggested baboon traits.

Hunting and meat eating by baboons was studied by Shirley Strum in a troop nicknamed the Pumphouse Gang at Gilgil in Kenya. Strum has now studied these animals for more than twenty-five years, but in the early 1970s she observed their hunting behavior in particular. Her study of meat eating showed that the Gilgil baboons ate meat more frequently—once per day—than any other known population of nonhuman primates. The exact pattern of both hunting and sharing varied depending on the individuals involved in each hunt, but the baboons at Gilgil were more sophisticated in their hunting behavior than any other baboons that have been studied. Strum reported cooperation, particularly when a particular adult male was involved in the hunt. This male appeared to be a catalyst who promoted hunting by other males, females, and even juveniles. The Glgil baboons thus provide a model for a cultural basis of hunting; they suggest that aside from any energetic concerns about the rate of return in hunting by baboons, meat eating is a learned tradition that may be exterminated or initiated depending on the composition of the group. Past experience, observation, and imitation of others dictated much of the meat-eating behavior of the Pumphouse Gang. Individuals did not scavenge carcasses unless they had some prior experience with the carcass or could watch another group member eating from it.[13] Why one population of baboons scavenges while another does not may be due to experience and learning opportunities. This hallmark of complex social behaviors, including hunting, certainly applies to chimpanzees as well. In the Taï forest of the Ivory Coast, the majority of adult males who did the hunting in Christophe Boesch's long-term study of chimpanzees recently died from an outbreak of the ebola virus. Learned traditions of hunting may have died with them. If there can be a silver lining in such a tragedy, it will be to observe whether and how the hunting tradition reemerge as a new crop of young male hunters matures without the benefit of observing their elders.

Meanwhile, in the forests of Latin America there is another, much smaller primate that also hunts and eats meat voraciously whenever it has the chance. This primate is a monkey much smaller than a baboon, and in addition to its carnivorous habits it is also the most adept tool user among the New World monkeys. The capuchins of the genus *Cebus*, found across South and Central America, hunt as avidly and as successfully as chimpanzees. They engage in relay chase hunts that resemble those reported for both baboons and chimps. They also employ tools more than any other monkey. Their brain-to-body size ratio is very high, as is true of chimpanzees. Studies of captives have shown capuchins to be active and strategic food sharers as well. It has only been in the last several years that any attention has been focused on their hunting prowess and sharing behaviors. This lack of attention is in turn due to the general lack of interest anthropologists have shown in the smaller-bodied Neotropical monkeys as models of human evolution.

Meat eating, tool use, and large relative brain size therefore occur in two distantly related primate groups—apes and New World monkeys—and meat eating and related behaviors are also known in an Old World monkey, the baboon. Is this a ran-

dom evolutionary convergence, or has natural selection driven the coevolution of these traits? These
animals are exemplars of how effective nonhuman
primates can be as hunters. They are just three of
two hundred primate species, so one might argue
that meat eating was not a fundamental factor in
the rise of the human species. The coincidence of
traits among these species, however, is striking.

OTHER ORIGINS

Finally, there is no good reason to limit ourselves
to primates when using living animals to reconstruct the origins of intelligence and complex social
behavior. Anthropologists study prosimians such as
bushbabies and lorises because they are our primate kin, but they are so distantly related that what
we are likely to learn from them about the origins of
human behavior is extremely limited. If an anthropologist were to turn to the cetaceans—the dolphins and whales—for answers about the origins of
humanness, he would risk ridicule. However, if we
are interested in how complex social systems arose,
then comparing how the Darwinian process has
molded unrelated highly intelligent species may be
more informative than comparing two more related
but fundamentally different animals.

It is hard to imagine mammals living in more different worlds than a dolphin and a chimpanzee. And yet these two creatures are representatives of the only two evolutionary lines of the billions that have ever lived to produce big-brained, socially sophisticated, and highly intelligent beings. Their social convergence may be due to ecological factors. Dietary and foraging constraints on the two species may be similar. For dolphins, the food resource is fish, which are unpredictable in occurrence and widely dispersed. This patchy resource promotes foraging in small, fluid social units, paralleling chimpanzee society, which is structured around the patchy distribution of their favored fruits in African forests. Male chimpanzees form long-lasting bonds based on kinship, and these coalitions attempt to control females in order to obtain matings. Male bottlenose dolphins in a landmark study in Shark Bay, Western Australia, have also been reported to form long-lasting alliances that cooperatively coerce females to mate with them. This cooperative behavior is also valuable in driving off predatory sharks, and possibly in finding schools of fish.[14] Both species are highly communicative, with vocalizations that show regional dialectical variation and a strongly learned basis. These parallel social adaptations suggest that the social complexity and brain size increase that we see in ourselves have their roots in the social and ecological environment in which our ancestors found themselves.

The Missing Link Is Neither

The idea of a missing link in human evolution is deeply instilled in all of us. No matter how complete the evolutionary sequence of human fossils becomes, there will always be those who demand a missing link. By definition, the missing link is the most recent common ancestor of both humans and great apes, which must have lived immediately before the split of these two lineages in the late Miocene or earliest Pliocene era. But in reality, there is no such thing as a missing link or even a most recent common ancestor. The concept of a common ancestor is just a metaphor for the enormously complex process of speciation that preceded the emergence of the hominids.

Consider the process by which two modern primate species might form. A vast river changes its course in the Amazon basin, splitting what was once a population of monkeys into two smaller populations. Over millennia, each population goes its own evolutionary way due to an accumulation of mutations that do not penetrate into the other, now disjunct population. The result will be multiple new populations that are genetically, morphologically, and behaviorally different enough that we consider them separate species. The criteria we use to recognize the populations as distinct species depends on which concept of species we use. Initially, there was not a single different population that gave rise to other, daughter populations. Rather, there were a number of populations, each with its own set of slight genetic and morphological and behavioral differences. We see this variation in geographically distinct populations today in many animal species. It is a gross oversimplification to imagine that there was one or even just a few populations of ape ancestors that gave rise to the hominids. In reality, there may have been numerous ones, and they may have differed substantially in morphology and behavior from one another, and had periodic contact during which their genes mixed. To invoke the missing link by calling any one of these populations the founder of our gene pool is almost certainly wrong. But because we are so limited in our knowledge of the fossil record, and even more in understanding what the fossil record can inform us about behavior, we employ the common ancestor logic in trying to model early hominid behavior.

Early Hominids as Weaklings

One of the most indelible images in our depiction of early humans is that they were weaklings. As humans became human, they lost two adaptations that characterize the great apes: climbing ability and long canine teeth. The usual depiction of these

nascent humans suggests that they were not particularly well adapted to either their new grassland
environment or to the forest habitat they are said to
have left. We see them traveling warily across the
savannah, armed with only their brains against the
carnivores that eagerly preyed on them. Often, their
gait is depicted as a shuffling, inefficient form of
semibipedalism, as though the creatures had some
half-formed ability to walk. This caricature is pervasive even among human evolutionary scientists—indeed, it was present in both a recent widely
watched documentary on human evolution and a
recent book about human nature. The image of
early humans taking up bipedalism after a long ancestry of ape quadrupedalism is implicitly that of
people who take up a sport late in life—that, because it is not their lifelong habit, they will take
time to do it well and may never be really proficient. This is nonsense, since at each stage of the
evolutionary process natural selection molded a
means of moving about that was efficient enough to
be favored and perpetuated. There was no way to
predict, nor any a priori reason to think, that evolutionary change in the way apes moved would ultimately produce an upright walker. We do not know
why hominids became bipedal, but we can explain
many of the consequences, such as greatly enhanced energetic efficiency over the ape's knuckle
walking.[15] Nevertheless, the shuffling protohominid

is a mainstay of popular and scientific accounts of early humans. The depiction of early humans as weaklings is also odd considering that we do not see apes in this light; what could be more gracefully agile than a chimpanzee? Even when chimpanzees travel on the ground, they are hardly defenseless, mobbing dangerous animals such as leopards and driving them away through joint effort.

Our ancestors may not have had huge canines, but they certainly may have been highly efficient killers and predators. During the years I conducted research at Gombe there was one elderly male chimpanzee, Evered, who was an accomplished hunter even though in his last years he had lost the muscle tone needed for treetop agility as well as nearly all of his teeth! Even modern people who lack anatomical adaptations to tree living but who live in forested environments have a tree-climbing ability far beyond that of anyone raised in Western society. Neither chimpanzees nor humans have any anatomical traits that specifically adapt them to a predatory way of life. Instead, both use their ability to hunt socially and cooperatively to compensate for a lack of such adaptations. One of the arguments against a hunting ancestry for early hominids has been that they lacked such adaptation and therefore were forced to scavenge for carcasses as their sole source of meat. This is a highly implausible scenario, in part because early hominids could make a living on the numerous small and medium-sized mammals with which they shared their habitat.

Our deep preconceptions influence and constrain the ways in which we theorize about the early nature of humanity. These preconceptions change, but they are always constrained by the limitations of our evidence and by the prevailing biases of the day. The extent to which these models reflect reality versus our own reflections of ourselves is embodied in some of the influential models of human origins that follow.

MAN THE HUNTER

In 1966 about fifty anthropologists who studied the life ways of traditional foraging people gathered in Chicago for a conference to examine the status of the world's hunter-gatherers.[16] Perhaps the foremost scientific conclusion that came out of the meeting was that the importance of meat in the diets of foraging people had been exaggerated. This was deeply ironic, since the most influential and ultimately notorious perspective to emerge from the meeting came to be known as "Man the Hunter." Sherwood Washburn, the most prominent and listened-to biological anthropologist of his day, and Chet Lancaster contributed a paper called "The Evolution of Hunting." It set out to explain how and why the human brain had experienced a 3.5-fold increase in size and complexity since the dawn of humanity. Washburn and Lancaster claimed that "our intellect, interests, emotions, and basic social life—all are evolutionary products of the success of the hunting adaptation"[17] (p. 293). They were referring mainly to our more recent human ancestors during the Pleistocene. In their view, hunting for game animals was at the soul of the human experience. However, they noted that the sexes had different roles in many traditional societies when it came to acquiring meat. Men, according to Washburn and Lancaster, hunt, while women gather. This view put men in the important role of obtaining the highest-quality nutrients and the calories that their households would use. Hunting requires communication and coordination of action among the hunters. This placed an evolutionary premium on intelligence and communicative ability in order to successfully track and hunt down potentially dangerous prey. Men did this, and women did not. Moreover, Washburn and Lancaster linked the deep human love of hunting to the equally deep human love of going to war and to acts of aggression in general. The fact that it is almost always males who carry out these acts served to reinforce the idea that men had a natural right to occupy the glamor role of clever-minded forager, meat provider, and conqueror in

human societies. Ever since, theories of human evolution have focused on male activities rather than
female as the core human adaptations.

In principle, the evolutionary logic on which Man
the Hunter was based was sound. It was an analogy
based on the behavior of traditional foragers that
employed a model of cognitive evolution equating
a fundamental change in human anatomy and behavior that was carried out by one sex only—men.
However, we know of many cases in which, due to
genetic linkages called "pleiotropic effects," one sex
exhibits traits that natural selection clearly produced in the other sex only. Male nipples are an obvious example. The serpentine neck of the giraffe,
long thought to be the product of natural selection favoring those giraffes that could reach high
branches to forage more effectively than short-
necked neighbors, was probably not the result of
natural selection for neck length. Male giraffes use
their necks to combat for females, twisting and
thumping each other in a dominance struggle.
Longer-necked males have greater mating success,
though at the same time they suffer greater mortality from predators than smaller males.[18] All else being equal, natural selection would favor shorter
necks. But not only male giraffes have long necks;
females do, too, because of pleiotropic effects. In the
same way, the early hominid brain could have

increased in both males and females even if only males were hunting for meat.

The intellectual stakes were higher in Man the Hunter, however, since it was about the roots of human gender relations, not giraffe necks. The response to Man the Hunter was angry and its impact long lasting. Many anthropologists were angered by the suggestion that a hallmark in our ancestry was effected by natural selection for male cognitive abilities, implying that women were merely carried along in some genetic linkage. Anthropologists Adrienne Zihlmann and Nancy Tanner pointed out that in some of the traditional societies that are most vaunted for the man's role in hunting, up to 85 percent of the animal protein obtained by a household came not from men at all, but from the less glamorous role of women gathering foods such as nuts, tubers, and small animals.[19] The role of human females had been neglected in Man the Hunter, according to the anthropological community, due to the gender politics of scientific advances and partly due to the failure to appreciate the role of women in foraging.

In the largest cross-cultural database that exists—a survey of 179 societies that examines how labor is divided in human groups—men alone hunt in 166, both men and women hunt in 13, and in not one do women alone do the hunting. Women, on the other hand, are the main gatherers of plant foods in about two-thirds of societies in the same survey.[20] So the reality of male predominance as hunters and of women as gatherers is not in dispute. Instead, anthropologists began to realize that although men hunt, they often fail to catch enough prey to sustain the family, and this task falls to women. Men might kill one giraffe and talk about it around the fire at night for a year until another is killed. In the reaction to Man the Hunter, the fact was lost that while meat may not be the *valuable* food resource it had been assumed to be, it is nevertheless the most *valued* food resource in most human groups, including among foraging people.

The backlash to Man the Hunter permeated all fields of anthropology. In the study of nonhuman primates, it contributed to a reappraisal of the way in which the field was practiced. Observers had always tended to focus more on the behavior of males than females, because they are often bolder and therefore more visible.[21] The practice of observational primatology was made more systematic when it was recognized that females also played a central role in the primate group. More recent theoretical advances made it clear that females rather than males are often the central players around which the mating system is structured.[22] Man the Hunter's backlash eventually led to an engendering of the field of archaeology; it was recognized that the role of women in early human societies had long been ignored in favor of the often more visible role of men. Stone tools made for butchering carcasses will preserve in the fossil record, while the implements of gathering made and used by women might not. Women nevertheless performed critical tasks and occupied spheres of influence in antiquity that went far beyond where the largely male archaeological community had considered them for decades.[23] Archaeologists today refer to the failure of an earlier generation of scholars to consider the role of women as a "Paleolithic glass ceiling."

A GOOD STORY

A good model of human origins should provide a number of hypotheses about our ancestors' behavior and anatomy that interweave in a sensible way. That is, the model must be both internally and externally consistent. It must explain the origins of those traits that are uniquely human above all else, since other traits that we share with the great apes are likely primitive ones that we possess simply due to a common ancestry. What is left after these are stripped away is those traits that define our humanness. To retrospectively build a human ancestor requires that we consider the roots of the key components. Some models make wonderful stories; like a novel they may be internally consistent even though they do not overall accord well with physical evidence. For example, the idea that humans passed through an aquatic phase in prehistory has been advocated and accepted in some popular accounts,[24] even though the specific lines of evidence in the model have no support.[25] It is simply an attractive, internally consistent story about who we are and how we came to be human.

The key adaptations that we must consider will vary depending on what stage of human evolutionary history we are thinking about. For instance, when imagining the common ancestor of all hominids, the key character is bipedalism, arising at least five million years ago and exhibited by no other primate. Our very large and complex brains, our tool-using capabilities, the increased amounts of meat in our diet, and our unusual social system all evolved at much later dates—2.5 million years for stone tool use and less than 200,000 years ago for a modern level of brain-size increase. The theories of hominid origins that have gained the most attention and notoriety have been those that have woven the greatest number of human traits together in an internally consistent way, even though we will see that these often become houses of cards by virtue of the number of variables they seek to link. In the following sections I discuss some key adaptations and how they fit into a portrait of early hominid behavior.

Key Adaptations

Bipedalism

No aspect of modern humans has been speculated about more than our unique bipedal posture and locomotion. Whatever the protohominid's mode of travel, bipedalism would probably have evolved only if it had increased the efficiency of movement in the emerging new species. For instance, Karen Steudel of the University of Wisconsin has analyzed the energetic efficiency of bipedalism in order to assess the hypothesis that early humans became more efficient travelers when they became bipedal; she disputes this widely held notion.[26] In the 1970s, studies of chimpanzee walking patterns led researchers to claim that bipedalism required no less an output of energy per unit of distance than quadrupedalism.[27] Later, anthropologists Peter Rodman and Henry McHenry found the opposite result; they reported that bipedalism must have arisen due to its far greater efficiency over the knuckle walking of apes.[28] Steudel argues that while bipedal walking is more efficient than knuckle walking, it is not likely to have been more efficient in its earliest incarnations than the quadrupedal alternatives available at that time. In other words, natural selection probably did not favor the continued evolution of bipedal locomotion in emerging hominids due to its efficiency unless our common ancestor with chimpanzees was a knuckle-walker. Ruling out improved energetics as the primary stimulus for bipedalism leaves a number of other reasons that have been promulgated. For example, here are some published explanations, along with at least one piece of contradictory information:

1. Being upright gives a height advantage to intimidate predators and other hominids.[29] *Problem*: Why is it important to be permanently upright? Standing upright for just a few seconds would achieve the same result.

2. Being upright allows an early grassland hominid to see over tall grass.[30] *Problem*: Same as above, plus the doubt over whether early hominid evolution really occurred in grassland versus woodland habitats.

3. Being upright reduces one's exposure to intense tropical sun and heat, thereby reducing heat stress on the savannah.[31] *Problem*: Again, the evidence that this key period of evolution took place on the savannah is now considered shaky.

4. Being upright is not about walking, but rather about posture when foraging. The bipedal posture may have evolved to allow apes to pull down low-hanging, fruit-laden branches,[32] or to allow for better tree-climbing ability on vertical trunks.[33] *Problem*: Neither of these receives strong support from the behavior of modern quadrupedal chimpanzees.

5. An upright walker has its hands freed for carrying food, offspring,[34] or tools.[35] I will deal with this last scenario below, for it incorporates some of the most widely held assumptions that have recently dogged models of human origins.

BIPEDALISM AND MEAT EATING

In 1981 physical anthropologist Owen Lovejoy published a paper that proved an influential but controversial model of human origins. He suggested that we should look at both bipedalism and the unusual reproductive system of humans to establish the likely social behavior of the earliest hominids as well as why they became bipedal. Unlike chimpanzees and bonobos, in which females possess large fluid-filled swellings when ovulating, human females conceal their ovulation. Lovejoy saw the roots of female manipulation of male behavior in the concealment of ovulation; unable to time the exact period of ovulation, male protohominids would have had to remain near the female in order to mateguard against the possibility of cuckoldry. At the same time, increased use of savannah habitat led to efficient bipedal walking. Lovejoy hypothesized that male provisioning of stay-at-home females resulted. Males used their newfound freedom of the hands to carry meat back to females. They thereby enhanced females' nutritional status, enabling an increased reproductive rate that was good for both sexes.

Unfortunately for Lovejoy's model, there is no reason to think that advertisement of ovulation has ever been a part of hominid biology. Chimpanzees and bonobos had a common ancestor with other

great apes, and probably evolved their swellings after their lineage had split with other apes. This is
the most parsimonious explanation and would
explain why none of the other apes or humans
possess swellings while both the chimpanzees and
bonobos do. The concealment of ovulation in women
is simply the retention of a widespread primitive
primate feature. The swelling is a chimpanzee-
bonobo feature that evolved sometime between 5
and 2.5 million years ago. Moreover, we have little
reason to assume that early hominids would have
been monogamous, any more than all modern human societies are (the majority are polygynous).
Some form of polygyny, the social system of three of
the four great apes and of the majority of human
societies, is the probable social organization of our
common ancestor with the apes. Lovejoy's model is
thus internally consistent but fails in the face of the
evidence.

Karen Steudel argues that bipedalism is not likely
to have replaced quadrupedalism on energetic
grounds alone.[36] So we might consider the existence
of an early bipedalism and a later bipedalism. This
is speculative since we as yet have no fossil that
exhibits a clearly different, unknown form of bipedal walking than ourselves.[37] However, it must have
existed, since bipedalism arose over thousands of
generations with each intermediate form molded
by natural selection serving some useful purpose

of its own. We do not know if the common ancestor of humans and great apes was a knuckle walker. Since all of the four great apes walk either on their knuckles or on the sides of their fists (orangutans), it is possible that there was a knuckle-walking stage in human evolution, as proposed many years ago by Sherwood Washburn and others.[38] If bipedalism arose for reasons other than enhancing walking efficiency, however, then the concern that it is more efficient than knuckle-walking but not more than quadrupedalism is unnecessary. But an early type of bipedalism, whether it arose from knuckle walking or not, could certainly have characterized the earliest hominids. The cause of this bipedalism could have been any of the hypotheses I presented earlier, or one that has yet to be proposed. The more modern form of bipedalism that we see in both ourselves and our known ancestors could have arisen later and been enhanced for energetic reasons.

One of the most striking differences between the foraging behavior of chimpanzees and that of humans is that chimpanzees crave the meat of other animals but do not search for it. Instead, they forage for plant foods and eat prey animals opportunistically in the course of looking for fruits and leaves. Certainly they are skilled hunters, particularly when in large groups. They could presumably obtain much larger amounts of meat if they actively searched for it. Yet there is little evidence that chimpanzees search for meat.[39] The only factor that could reasonably account for the chimpanzees' failure to search for food is that the return rate on their energy expenditure is not enough to do so. A biped can walk longer distances in search of desired food compared to a knuckle walker. Thus, if there was a knuckle-walking stage of hominid evolution through which early hominids passed, it would have precluded searches for unpredictable, moving sources of meat. Once bipedalism had evolved to a point of energetic efficiency, active searching could become justifiable, and meat would increase as a percentage of the diet.

Any model that seeks to explain human evolution must explain the enormous increase in the size of the neocortex of the brain in humans. I have already argued that, based on the behavior of chimpanzees, hominids ate meat much more frequently and at an earlier stage in their evolution than is commonly thought. This meat eating probably increased as early hominids began to use larger-sized animals as prey and as they began to use tools to make use of the carcasses. As the size of the brain increases during such a trend, one of the body's metabolically most expensive organs to maintain must be nourished. Where does the energy come from to nourish such a large brain? Leslie Aiello and Peter Wheeler have hypothesized that the energy to nourish an increasingly larger brain came from a metabolic trade-off in which the size and scope of the metabolic investment in the gut were reduced.[40] They suggest that in the course of primate evolution, the size of the brain has coevolved in inverse relation to the size of the digestive tract. This would partially explain why fruit-eating and meat-eating primates have shorter guts; these foods are more easily digested, allowing more energy to be devoted to increasing brain size. We know that leaf-eating animals sometimes have smaller brains and lower basal metabolic rates than fruit eaters of comparable body size. This relationship holds true among primates as well. Aiello and Wheeler suggest that the australopithecines improved the quality of food in their diet over that of their forerunners and thereby enabled a continued degree of encephalization. Likewise, among living primates the capuchin monkeys have among the highest brain-to-body-size ratio and also one of the highest-quality diets. This diet includes a high percentage of vertebrate animal protein. The implication of Aiello and Wheeler's "expensive tissue hypothesis" is that in a wide range of animal groups, those species that eat a high-quality diet should show greater brain development than those with a poor-quality diet. This has yet to be shown. But the notion that a high-quality diet frees the metabolism of an evolving hominid to develop a larger and larger brain is extremely appealing because it would explain both the trend toward greater encephalization and toward more meat in the diet of the evolution of the human lineage.

Ape Nature

I asserted—and I repeat—that a man
has no reason to be ashamed of
having an ape for his grandfather.

Thomas Huxley,
Letters (1900).

Much of what we know about the great apes today
would have been considered science fiction in 1960.
And many of the ideas that we are just now beginning to accept will be established facts for students
of great ape behavior in future decades. Tool manufacture and use, cooperative hunting and nepotistic
food sharing, Machiavellian social tactics, infanticide, sex for social bonding rather than reproduction, and intergroup warfare were once considered
exclusively human traits. All have been reported
among chimpanzees in the past three decades. In
this chapter I describe the societies of the four great
ape species, examining the threads of behavior and
ecology that unite this very diverse group. The
hunting and meat-eating behavior of chimpanzees

is a special focus, because this feature distinguishes them from the other apes and links them to the origins of our own meat-eating behavior. According to Darwinian principles of sexual selection, evolutionary biologists view each sex as having its own reproductive agenda but different ways to achieve reproductive success. What is good behavior for males may be unfavorable behavior for females, and vice versa. This intellectual framework has given evolutionary biologists a powerful tool for examining the social evolution of our close relatives.

Humans evolved from apelike ancestors, and, as already discussed in chapter 2, the ape and human lineages split about 6 millions years ago.[1] Subsequently, the ape lineage split as well, producing both the modern chimpanzee and the bonobo. Just prior to the ape-hominid split, the gorilla had followed its own evolutionary path, and still earlier the Asian and African ape lineages had diverged, yielding the modern orangutan. The four great apes are the living tips of rich evolutionary branchings that flourished between 10 and 20 million years ago. Nearly all the apes that have ever lived died out long ago. The remaining few are all the living evidence we will ever have of our evolutionary roots. In addition to the sparsity of living ape species, our ignorance is compounded by the dearth of African ape fossils. While the orangutan's ancestors may be represented in the fossil record,[2] no one has yet discovered any fossil evidence of what appears to be the immediate ancestor of a gorilla, chimpanzee, or bonobo. While we should not assume that chimpanzees and bonobos have remained unchanged over the millions of years of evolution that separate them from us, we can assume that they are more similar to our shared ancestors than modern people are (table 3.1).

Information about the great apes has emerged slowly. This is because all the great apes live long lives, passing through the same developmental stages at about the same age as humans. Moreover, apes have slower reproductive rates than any human populations, in some cases dramatically so. Chimpanzee mothers give birth only every five years, and female orangutans produce a baby only every eight years. This means that studying several ape generations requires at least 2–3 human generations of researchers; forty years of observation in Gombe National Park, Tanzania, have only compiled information on just over three generations of chimpanzees. And learning about the social behavior of an unsociable ape like the orangutan takes even longer, simply because the opportunities to observe crucial social interactions arise infrequently.

TABLE 3.1. Overview of the societies and ecologies of the living great apes.

Species	Social System	Dispersing Sex	Female-Female/ Male-Male Bonds	Diet	Meat in Diet?
Chimpanzee	Fission-fusion polygyny	Females	Weak/ Strong	>75% fruit	Small mammals eaten regularly
Bonobo	Fission-fusion polygyny	Females	Strong/ Strong	>75% fruit	Small mammals eaten occasionally
Gorilla (western lowland)*	Multimale group polygyny	Females	Weak/ Weak	45–70% fruit	None
Gorilla (mountain)	1 and multi-male group polygyny	Females	Weak/ Weak	<5% fruit	None
Orangutan	Mainly solitary	Females (?)	None/ None (?)	>70% fruit	Small mammals eaten occasionally

*Lowland and mountain gorillas are presented separately because of their differences and because molecular evidence suggests they should be considered separate species. The same may be true for eastern and western chimpanzees and Sumatran and Bornean orangutans.

Chimpanzees

An early September morning in Gombe National Park in western Tanzania. A party of chimpanzees is on the move, traveling up a steep mountain slope toward a stand of fruit trees. The group of ten animals had nested in the same trees the night before. As they approach the crest of the ridge and reach the ripe fruit, they begin to pant-hoot, the loud calls reverberating off the valley walls. These calls are answered from the valley below by another party also headed this way. The party contains four adult males plus females and their infants. One female carries a huge pink sexual swelling, a billboard of her sexual state. After twenty minutes of gorging on the fruits, the second party arrives, containing five more adult males including Wilkie, the current top-ranking alpha. An excited greeting display follows between the males, and Wilkie charges time and again up and down the slope dragging branches, his hair bristling as he chases other lower-ranking males out of his path. Meanwhile, another high-ranking male approaches the swollen female, a recently arrived immigrant from the adjacent community. When she does not immediately present herself for mating, the male attacks her, chasing her to the top of a tree where they mate. Later, some members of the combined party depart, leaving a new mixture of chimpanzees behind. The parties continue to depart and recombine, making and remaking a jigsaw puzzle of associations throughout the day.

Our understanding of chimpanzee society has evolved very slowly because of the difficulty in obtaining a clear portrait of their complex mating system. During the early 1960s it was widely held that chimpanzees, unlike other group-living primates that had been studied, had no group structure whatever and that relations among individuals were constantly changing. Jane Goodall had been conducting research in Gombe since 1960 and had pioneered the observational study of wild chimpanzees, discovering meat eating and tool use in the process. Working in the Mahale Mountains of Tanzania some 100 kilometers south of Gombe, the Japanese primatologist Toshisada Nishida described a chimpanzee society based on the "unit-group" (later called the "community" by Western primatologists).[3] The community is a local breeding population of anywhere from twenty to over one hundred chimpanzees; it occupies an area that is defended against intruders and has a stable membership within which there are no consistent grouping patterns except mothers and their immature offspring. The community members come together and depart unpredictably all day long. This complex sort of primate society is called fission-fusion, and it serves an ecological purpose for its members. The smallest irreducible unit of chimpanzee society is mothers and their infants, and foraging parties form to provide females with optimal fruit foraging opportunities. Small parties allow females to maximize their food intake by avoiding competition for fruit from other chimpanzees. Females who are sexually receptive seek out males or are sought by them to form larger parties. Party size in chimpanzees is therefore a function of both food patch size and the presence of sexually receptive females.

It was learned through the slow accumulation of life histories that female chimpanzees, upon reaching puberty at about age twelve, usually migrate from the community of their birth into other communities to take up life as resident adult females. These females, when sexually swollen, mate with multiple males. The view that chimpanzee social systems are characterized by casual promiscuity thereby became entrenched, in spite of Goodall's early observations of intensely aggressive competition among males for estrous females. Maturing males, meanwhile, remain in their native land for life, bonding with other males to whom they are often related. Male chimpanzees tend to socialize with one another, and these alliances patrol the community's territorial borders and try to control females as well. At both Gombe and Mahale, lethal territoriality between neighboring communities, often called warfare (though the observed pattern was more a series of commando raids into neighboring communities), reinforced the view that chimpanzee society is male controlled from both

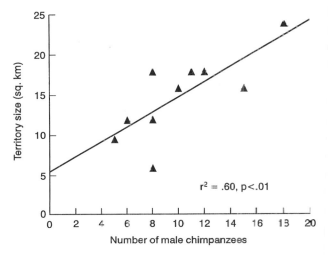

Fig. 3.1. Among Gombe chimpanzees, the size of the community's territory has varied in relation to the number of adult and adolescent males from the 1960s through the mid-1990s.

within and without. The territorial patrols attack neighbors and sometimes acquire new females; females are essentially reproductive chattel over which males compete. In eras in Gombe National Park when the number of males has been greatest, their community territory has also been largest, attesting to the importance of male defense to territory size (fig. 3.1). The majority of hunting by chimpanzees is also by males. Within the community, males are dominant to females and at times behave quite brutally in their subordination of them.

Fig. 3.2. Adult chimpanzee (*Pan troglodytes*) Frodo, Gombe National Park, Tanzania. (Photo by Craig Stanford)

Chimpanzees also live, however, in a society composed of females who actively choose their mates by inciting male competition when they are sexually receptive (advertised by the presence of fluid-filled swellings on their genital regions), and who with their infants form the nuclear units of the social system. These females forage independently from males to minimize competition for food and socialize with the males mainly when it suits their own reproductive interests. And although males may appear to control female reproductive behavior, the promiscuous, mate-soliciting female chimpanzee is really the sex that is in charge of the mating system. Pascal Gagneux and colleagues recently studied the mitochondrial DNA of chimpanzees in the Taï forest, and found that half of the babies born were fathered by males from outside the community. This startling finding suggests that some females had disappeared from the community for as little as forty-eight hours while they were swollen, apparently to seek out males from neighboring communities to sire their offspring.[4]

When comparing human societies, anthropologists speak of cultural variation. The diversity of chimpanzee behavior has become clear as studies from different regions of Africa document various styles of tool use, hunting, and social behavior. In addition to Gombe and Mahale, a third long-term study in the Taï National Park in the Ivory Coast and a fourth in Kibale National Park in Uganda have produced new perspectives on tool use and hunting. While chimpanzees in Gombe National Park use fishing poles modified from twigs to lure nutritious termites from their mounds, they ignore

FIG. 3.3. Chimpanzee tool use: Gigi fishes for termites in Gombe National Park, Tanzania. (Photo by Craig Stanford)

the rocks strewn across the forest floor that would be useful hammers for cracking hard-shelled fruits. Two thousand kilometers to the west, the chimpanzees of Taï National Park in the Ivory Coast, a rain forest poor in loose rocks, gather stones to use as hammers. They place nuts into depressions on tree roots, and bring the stone-and-wood hammers down onto this anvil to get at the nut's flesh. Availability is not the mother of invention, though: Taï chimpanzees do not use fishing probes to get the termites in their forest, nor do Gombe chimpanzees use the abundant stones in their forest as hammers.

These appear to be learned traditions similar to those passed on in traditional human societies. Losing even one chimpanzee population to extinction thus means the permanent loss of any traditions that might have been unique to a particular chimpanzee culture. Just as an anthropologist walking from one village to another in the Amazon basin finds tribal groups with different languages and rituals, a primatologist traveling only a hundred kilometers from one study site to the next can encounter a new chimpanzee culture with features different from all other chimpanzee cultures.

Hunting

September 1994. In a shady ravine in Gombe National Park, a party of chimpanzees is eating a kill of colobus monkeys. Prof and Frodo each has a small colobus monkey, and the crunching of bone and ripping of skin can be heard from where I sit, five meters below on a leaf-covered hillside. The captors sit on several adjacent branches while the allies and relatives sit a few feet away, staring jealously and occasionally approaching to beg meat with a supplicatory extended hand. A low-ranking male, Beethoven, approaches Prof and begs for a scrap by placing his fingers at Prof's lips; he leaves them there as Prof ignores him. Finally Prof turns his back on Beethoven, who adjusts his position to continue his silent pleas. Eventually, Beethoven is rewarded for his

efforts with one leg of the colobus, perhaps surrendered
by Prof to receive a few minutes of peace while he eats.
Frodo meanwhile has no intention of sharing with any-
one other than his mother, Fifi, and her son, Frodo's lit-
tle brother Faustino. The sharing of meat goes on for
almost 90 minutes, and all the while most of the hunt-
ing party sits patiently hoping that some small bits of
the catch will fall from the diners' hands and spatter the
leafy ground below, where they will scavenge a few
grams of protein.

In the world of primate behavior, the description
by Alfred Lord Tennyson of the struggle for existence as "Nature, red in tooth and claw" is nowhere
more evident than in the relationship between the
chimpanzee and its prey. Chimpanzees eat mainly
ripe fruit, leaves, flowers, seeds, plus some insects
and other invertebrates. They are also efficient
predators that in some forests kill hundreds of other
mammals per year. Chimpanzees are the only great
ape to avidly hunt and consume large quantities of
meat. They clearly relish meat. After a kill, the
younger members of the hunting party will sometimes sit below the tree limb on which the hunters
are crunching on bones and tearing flesh, hoping
for scraps that fall to the ground. After an hour's
wait, these scroungers may receive a few drops of
blood or shards of bone.

Unlike other nonhuman primates, the capture of

FIG. 3.4. Meat eating by chimpanzees: Freud with red colobus carcass, Gombe National Park, Tanzania. (Photo by Craig Stanford)

meat may entail cooperation. A highly ritualized sharing of meat occurs after a kill is made. When a lion kills a zebra, the drama of the hunt ends the moment the zebra is killed; a moment after its death the zebra carcass is just lunch. When a party of chimpanzees makes a kill, however, there often ensues a social frenzy. At one level this is simply about getting a scrap of meat, for which members of the hunting party will beg, borrow, or steal. But underlying the nutritional aspect of getting meat, part of the social fabric of the community is revealed in the dominance displays, the tolerated theft, and the bartered meat for sexual access. The end of the hunt is often only the beginning of a whole other arena of social interaction.

Although meat is a small component of the overall diet, its role in some chimpanzee societies is disproportionately great. What chimpanzees eat when they eat "meat" is the muscle tissue, brains, bone, hair and, viscera of captured animals. The upper limit on the body size of the prey is only about 15 kilograms (30 pounds), but the combined weight of prey may be much more, since monkeys and other prey are sometimes caught in groups.[5] Chimpanzees also steal the carcasses of prey caught by other animals such as baboons, and on rare occasions they will scavenge dead carcasses found on the forest floor while they are foraging for plant foods. But the vast majority of prey is caught by the hunters themselves.

There is little evidence that hunting is undertaken with a prior search image or intent to hunt in mind. Instead, chimpanzee parties spend their days traveling the forest in search of ripening fruit and other plant foods. When they encounter prey animals they sometimes attempt to capture them. In all of the studies in which chimpanzees have been watched for years on end, the major prey they have been seen to eat is the red colobus monkey, a long-tailed treetop monkey found in various forms in forests across equatorial Africa. In addition, chimpanzees take piglets of the common wild pig spe-

cies found in many African forests, fawns of bushbuck antelope, and a variety of other monkey species. The total quantity of meat eaten in a year,
while small compared to most human societies, is
surprisingly large considering that chimpanzees
were formerly thought to be complete herbivores.
In peak hunting years, the Kasakela chimpanzee
community at Gombe consumes nearly a ton of
meat, and the contribution of animal protein to the
diet can rival that of some human societies.

The hunt itself is heart-stopping action, usually
carried out on a group basis. The reward for making a kill is a package of meat and viscera that is a
richer package of protein and animal fat than anything else chimpanzees could find in their habitat.
Red colobus monkeys in Gombe live in large groups
with many males, and these males actively defend
their females and young with vicious counterat-
tacks on their would-be predators. As a party of
chimpanzees travels along a trail in the forest, they
come upon a group of colobus feeding overhead.
Colobus mothers gather up their infants while the
group's males position themselves between the
chimpanzees on the ground below and the rest of
the group. If the chimpanzees decide to hunt—often one or two hunters act as catalysts without
whom the predation opportunity would be passed
by—they begin to climb toward the intended prey.
The sounds of high-pitched colobus alarm calls and

excited chimpanzee screams fill the air as the hunters attempt to get close to their quarry. The hunters often target monkeys carrying small infants and will take the defenseless baby while leaving its mother unharmed. A hunt can last from a few minutes to more than an hour, and if it is successful from one to seven colobus may be killed.

Chimpanzee hunting behavior tells us much about chimpanzee society. Both females and males love meat, but only males hunt frequently. At Gombe, hunting occurs throughout the year, but is much more frequent in the dry season and during periodic "binges." Chimpanzees usually hunt in social groups, and the larger the hunting party, the greater their odds of making a kill. Larger hunting parties are more successful than solitary hunters or small parties. The increased chances of hunting success provide an incentive for joint action and some degree of cooperation during a hunt, though whether hunts are highly coordinated appears to differ from one chimpanzee society to the next.

In Taï National Park, the Swiss primatologist Christophe Boesch has reported extraordinary levels of cooperation among hunters. Some hunters act as drivers, pushing the colobus through the treetops toward other chimpanzees who have climbed into the monkeys' path to intercept them. Those hunters who fail to make a kill but who have contributed to the overall success of the hunt end up receiving a share of meat from the captor. In this system of altruistic reciprocity—help me catch a monkey now and I will reward you with a scrap of meat—Boesch sees the roots of sophisticated levels of cooperation in humans.[6]

At Gombe, by contrast, little overt cooperation of the Taï variety has ever been reported. Gombe chimpanzees hunt the way that many other social hunting animals do; everyone is intent on the same goal, and the sheer number of hunters trying to accomplish the same thing makes the odds that someone succeeds more likely. Before we judge the failure of Gombe chimpanzees to cooperate to indicate a lower state of social evolution, consider the two societies' styles. Gombe chimpanzees capture meat and share it nepotistically—with close relatives and allies—rather than with their hunting comrades. They also use the meat as a political tool of social manipulation, rewarding allies and snubbing rivals in plain view of those other members of the community to whom they may want to send a message. Toshisada Nishida and his colleagues have found the same sort of nepotistic use of meat among the chimpanzees of Mahale National Park.[7] Why should we think that cooperation is a more highly evolved art form than selfish manipulation? In our own social lives, each type of strategy has its purposes and rewards, and for both politicians and team athletes, different types of tasks require different tactics.

At Gombe, male chimpanzees who have meat will sometimes offer bits to swollen females, in exchange for which they receive matings. These sexual manipulations show both the males' ability to exert control over female reproductive behavior and possibly females' abilities to obtain meat without needing to kill it themselves.

Female chimpanzees are much less involved in hunts despite relishing meat. At Gombe, fewer than 10 percent of all kills have been made by females over the past twenty years.[8] Hunting carries the risk of injury from colobus monkey teeth; hunters sometimes come away from hunts with puncture wounds on their bodies. Exposing their offspring to a colobus's canine teeth during a hunt is a risk that females with offspring should try to avoid. Gigi, a female at Gombe not burdened by dependent infants because of her presumed infertility, was the only female actively involved in many hunts and a frequent killer of colobus during my research there.[9]

THE FOUR GREAT APES

BONOBOS

Stark contrasts but also very close similarities are evident between chimpanzees and bonobos (table 3.2). Even an armchair observer of ape behavior has heard about the hunting and warring tendencies of chimpanzees. For bonobos, the behaviors on which

TABLE 3.2. Chimpanzee and bonobo social systems compared: myth and reality.

Behavioral Trait	Chimpanzees (4–11 in party)	Bonobos (5–22 in party)
Sexual dimorphism	Males 10–15% heavier	Males 10–15% heavier
Male-male relations	Affiliative	Affiliative
Female-female relations	Weakly affiliative	Strongly affiliative
Male copulation frequency, Mahale vs. Wamba	0.20–0.29 times/hour	0.10–0.21 times per hour
Duration of maximum sexual swelling	13 days of 37-day cycle (35%)	14–15 days of 42-day cycle (approximately 33%)
Percentage of copulations when female is swollen	Nearly 100%	About 95%
Sexual behavior with same sex	Very rare	Frequent
Interbirth interval	5 years	5 years
Dominance	All adult males dominant to all adult females	Males dominant but defer to females at food sources
Intercommunity relations	Aggressive, occasionally lethal	Sometimes aggressive, sometimes not
Meat eating	Frequent, meat controlled by male	Rare (?), meat sometimes controlled by females

popular accounts focus is their lack of aggressive-
ness and richly sexual social lives. These contrasts
have received more attention than the far greater
similarities, and there is a widespread perception
that these two species represent a yin and yang of
humanity—a dualistic scenario of human nature's
roots. If chimpanzees, with their male dominance
machinations, their hunting, and their lethal territorial incursions, represent the darker side of human nature, then bonobos are often invoked as the
brighter side. At least this is the way many accounts
have portrayed the two species.

The two species have probably been on separate
evolutionary paths only for the past two and a half
million years. Before the human-created problems
of poaching and deforestation intervened, chimpanzees were an extraordinarily successful species
that had colonized a wide range of habitats across
the whole of equatorial Africa. They still occur from
dry woodland/savanna to primary lowland rain
forest, from sea level to over 2,500-meter elevation.
Bonobos, meanwhile, are limited to an expanse of
lowland rain forest in the Congo basin south of the
Congo River, where their numbers are precariously
low and dropping rapidly.

For decades, bonobos were the most poorly
understood of the great apes. In the 1970s, studies
began to point out shared traits of the "pygmy
chimpanzee" (as bonobos were then known), the

FIG 3.5. Bonobo (*Pan paniscus*). (Photo by Craig Stanford)

chimpanzee, and humans, and some scholars began to make the claim that bonobos served as the best analog for earliest hominid anatomy and behavior. This bonobo model was met with skepticism in many quarters, since the traits said to link humans and bonobos may equally reflect the 6 million years of bonobo evolution separate from its last common ancestor with humanity. The bonobo model also raised the issue of which of two equally closely related great apes to look to for insights about our roots.

Among bonobos, females form alliances to gain and share power against males. Females of one community have even been seen mating with the males of another in the presence of the females' own resident males.[10] This would be unthinkable in chimpanzee society. Bonobo society is a variation on the chimpanzee fission-fusion theme, and, like chimpanzees, male bonobos remain on their natal land. Both female chimpanzees and female bonobos emigrate from the natal community at or after puberty and eventually settle into lives in neighboring communities, leaving mainly unrelated females to live side by side. But bonobo communities do not live in completely peaceful coexistence, as half of intercommunity encounters involve aggressive behaviors such as chasing. No one has yet witnessed a lethal territorial encounter among bonobos, but they may occur; the total number of hours that field researchers have spent watching bonobos is a tiny fraction of the amount of time we have been watching chimpanzees.

Sexuality among bonobos is the most talked about aspect of their social life. Among both chimpanzees and bonobos, adult and adolescent females vividly advertise their fertility for a portion of the monthly menstrual cycle. This advertisement is the sexual swelling, the pink, fluid-filled anogenital sack that inflates for several days on each side at the time of ovulation. It is a billboard of sexual availability, one that excites males and incites competition among them, and may enable a female to size up which male or males are the most desirable. As an aspect of their hypersexuality relative to chimpanzees, female bonobos exhibit a sexual swelling for about two weeks of their six-week cycle. Female chimpanzees swell for about ten days of a shorter five-week cycle. Although chimpanzee births do not occur with any seasonality, swelling cycles are seasonal, peaking in the dry season in western Tanzania and influencing the pattern of party aggregations. Although the swelling does indicate an ovulatory state, female chimpanzees and bonobos also experience swellings while they are pregnant or lactating, though these nonovulatory swellings are less regular in their frequency and duration. When female chimpanzees or bonobos are swollen, they become highly attractive magnets around which males aggregate, and the females themselves become more sociable as well. The result is large parties that form and travel together, dispersing once the swelling subsides. Among chimpanzees, when "popular" females are swollen there is a sexual frenzy in the community as the alpha male attempts to preserve a sexual monopoly while the female herself actively seeks out other males, all of whom are interested and may challenge the dominant male for the opportunity to mate with her.

In 1991 I followed the male chimpanzees at Gombe during a week when the high-ranking female Gremlin was swollen and ovulating (we assume she was ovulating since she gave birth eight months later). Her behavior and that of the males belied the belief that chimpanzees are openly promiscuous. For six straight days and nights, alpha Wilkie sequestered Gremlin as they traveled together through the treetops in Kakombe valley. They were followed on the ground from dawn until dusk by an entourage of the other adult males, who stationed themselves at the base of whatever tree the mating couple were in and sat, often with erect penises, staring hopefully into the tree crown. If Wilkie allowed Gremlin to stray more than a few meters from him, a male from below would scramble up the tree, only to be driven back by an angry attack by Wilkie. During this time, only Wilkie's close ally Prof was allowed to approach and even occasionally to mate with Gremlin; she was otherwise off limits to all the suitors in waiting. This level of male possessiveness of females is not seen in bonobo society, mainly because females do not allow themselves to be controlled by males to the same degree as chimpanzees.

A premise about bonobo sexuality, and of the bonobo's relevance to understanding human behavior, is that bonobos and human females are the only primates who are sexually active outside the time of ovulation. Wild female chimpanzees rarely mate when they are not swollen, though in the confined boredom of zoos they may. Bold statements about bonobo hypersexuality, complete with erotic descriptions of their couplings, can be found in both scholarly publications and in the popular media. This sexual behavior is not just about conceiving babies. It is also thought to play a key role in social communication, as it also occurs between members of the same sex. Two female bonobos may enter the same tree to feed and rub their genitals together, apparently as a means of reducing what would otherwise be an unacceptable level of social tension between them. Nonreproductive sex clearly plays a strong role in human social bonding and social communication, and is made possible by the constant sexual receptivity that characterizes humans. Are bonobo females truly in the same class as humans in terms of nonprocreational sex? The most detailed study of sexual receptivity in wild female bonobos refutes much of the strength of this claim. Takeshi Furuichi found that although female bonobos do copulate occasionally when not cycling, most copulations (more than 90 percent) occurred when the females were maximally swollen, or nearly so.[11] Bonobo females swell for the same percentage of their menstrual cycles as female chimpanzees, but the bonobo swellings are longer in the raw number of days they remain maximally swollen.[12] This is only slightly less often than chimpanzees. Female bonobos thus appear to be somewhat more flexible in the timing of sexual receptivity than are other nonhumans, but they are by no means released from estrous cycles in the way that humans are.

Bonobos are also often portrayed as hypersexual in the frequency of their copulations, but mating frequencies overlap between the two chimpanzee species. Female chimpanzees copulate with numerous males during their fully swollen period; as many as fifty copulation bouts with eight males in a day have been recorded, and swollen females may copulate with several adult males in a five-minute period.[13] Bonobo females actively solicit sex from a range of males and may copulate multiple times per hour while they bear sexual swellings.[14] Overall, bonobos do not copulate more often than chimpanzees; the mean hourly copulation rates for the most successful male bonobos at one field site were approximately equal to those of the least successful male Gombe chimpanzees.[15] Even those bonobo researchers who have reported high mating frequency have acknowledged that this may be due in part to captive confinement.[16]

While female bonobos may not be truly hypersexual compared to chimpanzees, there are some fundamental differences between the two ape societies. One difference involves the presence of swollen females in foraging parties. Among Gombe chimpanzees, at least one female chimpanzee in the community is swollen about 50 percent of the time.

Therefore, only a small percentage of traveling parties contain a swollen female. By contrast, virtually every mixed-sex bonobo party has at least one sexually swollen female.[17] This is a very important difference between the two societies: access to reproductively active females is much greater for bonobo males than it is for chimpanzee males. This may in turn account for the lower levels of aggression among male bonobos than those reported for male chimpanzees.

Perhaps the most fundamental difference in the behavior of bonobos and chimpanzees is the relationship among females. Chimpanzee society is male dominated; adolescent males rise in rank by first dominating each adult female, and then they enter the bottom of the male hierarchy. Females, because they arrive as immigrants unrelated to other resident females, receive little support from the other females in their community; they are like sisters-in-law in a patrilocal household. Female chimpanzees also spend their lives alone much of the time rather than as part of a cohesive group.

Female bonobos also transfer to new communities at adolescence and also may not be immediately welcomed by females in the new community. But much of the similarity to chimpanzees ends here; bonobo females become partners in power, and alliances of females form that in some cases allow them to dominate males. The power base which in chimpanzee society rests solidly with adult males is therefore more female-centered in bonobos. However, even this difference between the two ape species may be smaller than previously thought, as female dominance is expressed mainly in feeding situations, in which males defer to females by allowing them to enter good feeding areas first.[18] In exchange for their deference, males receive sex. Thus, what may appear to be female-dominated feeding is in fact self-serving deference by males. In all other social arenas, male bonobos are dominant to females. Both Frans de Waal and Amy Parish have described patterns of female dominance, but again in zoo settings. As in other areas of social behavior, the stark contrast often drawn between these two apes of male versus female dominance may be a false dichotomy. If one considers dominance in the social arena rather than priority of feeding access, the pattern of dominance in bonobos begins strongly to resemble that of chimpanzees.

GORILLAS

A chilly, wet morning in the Impenetrable Forest of southwestern Uganda. A gorilla group is on the move. The silverback, his females and their babies, and a juvenile blackback male wake up in a small clearing, one that they created the night before. Due to the sheer size and

*weight of their bodies, large patches of the vegetation in
a forty-square-meter area has been flattened. The blackback watches as the silverback awakes and pushes his
way uphill out of the night nest area. The rest of the
group follows, and several minutes later they have all
settled into a patch of soft-leafed plants in a sheer hillside thicket. The gorillas are invisible in the thick
growth, but one can follow their movements by swaying
stalks and by the constant grunting, gurgling calls they
give as they forage: gorilla conversation. This, combined
with the growling of their stomachs as they eat their
breakfast, makes for a noisy scene. Then the silverback
and two females begin to climb. A minute later they are
thirty meters up on a moss-coated tree limb, foraging for
some epiphytic plants among the hanging ferns and or-
chids. The 200-kilo (400-pound) silverback feeds awhile,
then descends, bearlike and rump-first, down the trunk*

Gorillas have been very badly misunderstood in
the history of primate study. They were once thought
to be savage killers that would attack unprovoked,
a caricature dispelled only in the 1960s by George
Schaller's fieldwork. Schaller found gorillas to be
quite timid and tranquil in the presence of people,
not irascible. Following Schaller, work by Dian Fossey and her students on a population of mountain
gorillas living in the Virunga volcanoes on the
Rwanda-Congo border documented gorilla lives
in great detail. We learned that gorillas were foli-

Fɪɢ. 3.6. Mountain gorilla (*Gorilla g. Beringei*), Bwindi-Impenetrable National Park, Uganda. (Photo by Craig Stanford)

vores—leaf eaters that spent their day pushing, bulldozer-like, through impenetrable thickets of wild celery and nettles. They were considered the cows of the primate order, eating to live and living to eat, and able to subsist on browsing a poor diet by virtue of the serpentine digestive systems that curl within their massive bodies. The usual fare might be poor quality, but by the time it passes through the long gorilla gut it is broken down to the max. Again, cows. Gorillas seemed to support the rule that large-bodied primates are likely to eat a poor quality diet while living a low-energy lifestyle, while smaller primates eat a higher-energy diet such as fruit and have a more active lifestyle. In what little time they had left over after all the food finding, eating, and digesting, gorillas seemed to engage in social behavior, most often living in groups composed of one silverback plus his harem of females and their young.[19]

Recent fieldwork conducted elsewhere in Africa has shown that the sedentary leaf-eater portrait of the gorilla that came out of the early days of research, based on that one montane gorilla population, is a wrong characterization of the species. Mountain gorilla behavior does not closely resemble that of its lowland cousins, because the montane form is a remnant population living in a habitat utterly unlike that in which most gorillas evolved and currently inhabit. Throughout the towering primary rain forests, the swamp forests, and even in degraded secondary forest patches in Central and West Africa live most of the world's gorillas. This is the lowland version of the species, and its behavior and ecology are different. Far from the lethargic celery eaters of the mountains, these rain forest gorillas eat large quantities of fruit, and they forage far and wide to find it. Studies led by Caroline Tutin in Gabon, by Jorge Sabater Pi in Rio Muni, and by Juichi Yamagiwa in easternmost Congo showed that gorillas eat diets that are rich in fruit whenever and wherever they can. Recent studies by Melissa Remis and Michele Goldsmith have shown that they may even disperse into smaller subgroups, perhaps to locate patchy food sources and then devour them without competition. Moreover, these lowlanders are not reluctant to climb forty meters into the crown of a tall tree to find desired foods. In the remote Ndoki forest of the Congo, a team of researchers studying the relationship between gorillas and chimpanzees in the same forest have seen the two apes clambering about in the same trees together.[20]

Unlike the constant unpredictability of chimpanzee and bonobo groupings, gorillas live in more cohesive societies. Gorilla society consists of a small number of adult males to whom are bonded a number of females and their young. The number of adult silverback males is usually small, but up to six silverbacks have been reported to live in the same group. Because mountain gorilla silverbacks were often the victims of grisly poaching deaths during the 1960s and 1970s, we cannot be sure whether more than one silverback normally heads a group in times of lower silverback mortality. In any case, these males are the glue of the social group; when a male dies his females generally go their separate ways. The females have the same sister-in-law relationships that we saw in chimpanzees, and sisters may migrate together and end up in the same new group. These maternally related animals tend to support one another in squabbles, whereas unrelated females would not. Females are far smaller than and subordinate to adult males. But in spite of all appearances, females exert a key influence in the dynamics of gorilla group life; they decide which social group they prefer to belong to, and transfer among groups accordingly. These transfers most often take place following an encounter with another gorilla group. Since gorilla groups do not defend territories, they have home areas that overlap broadly like Olympics rings, and encounters between groups occur often in these overlap zones. The silverbacks square off and defend their females, and it appears that the females meanwhile size up new potential mates in the opposing group. This is a good example of the fallacy of the harem concept; the male may be larger and more powerful, but females are still active strategists in search of their own reproductive ends. Both females and males may emigrate from the groups of their birth. Those males that emigrate rarely enter another established gorilla group. Instead, they wander alone or bond with other bachelor males in search of eager-to-transfer females. When a group loses its silverback, the females left behind stand a grave risk of losing their babies due to infanticide committed by other

males. Death at the hands and teeth of another gorilla is one of the leading causes of death for baby
mountain gorillas.[21]

Orangutans

Imagine an extraterrestrial primatologist coming
to Earth to conduct a field study of the human species. How long would she have to follow several of
us around before she could confidently report back
to her planet a thorough account of human social
behavior? Unable to interview her primitive subjects and working alone in a strange setting, she
would have to select a small group of people who
represented all age and sex categories and then follow them closely, observing their every move in order to reveal the inner workings of human social
dynamics. This task would take decades. But it can
be done because humans in traditional societies
nearly always live in small groups that include
men, women, and children, so our exoprimatologist
could learn about the lives of all ages and sexes si-
multaneously. Now imagine the dilemma this scien-
tist would face if humans lived mainly solitary
lives. The length of time she would have to invest in
a study of social behavior would be many times
greater; she would observe lone individuals for
endless days in hopes of seeing what they do during their brief social gatherings. Even in a solitary

Fɪɢ. 3.7. Orangutan (*Pongo pygmaeus*). (Photo by Craig Starford)

species, the secrets of its life lie in the social arena. This is the case of the orangutan, and of the history of orangutan research.

We understand much about the lives of many individual orangutans, somewhat less about the relationships among individuals, and very little about the deep structure of orangutan society. Like gorillas, they exhibit extreme sexual dimorphism; like chimpanzees and bonobos, they are largely fruit eaters; like many primate females, they are choosy about which males they will mate with; and like all apes, they reach maturity and reproduce themselves very slowly. But unlike other apes—indeed all other higher primates—orangutans are mainly solitary, associating with others mainly to mate. Female orangutans hold territories on which they forage, their dependent offspring in tow. Their geographic distribution, on the Indonesian islands of Sumatra and Borneo, is mainly tropical rain forest, and it holds all the fruits and other foods needed by the mother orangutan and her young. An adult male who is twice her size, meanwhile, maintains a much larger territory that encompasses the ranges of several females.

Adult male orangutans come in more than one size. Resident males, who attempt to monopolize a number of females, tend to be large and use long-distance calls to drive away intruding males, attacking their rivals on sight. But their size also means that they are slow, and the large area conscribed by their territory cannot be effectively monitored for all of an individual's females at all times. This leaves the door open for transient males. These floaters were long thought to be adolescents, which was assumed to explain their smaller size and lack of the fleshy facial appendages that characterize full adults. But years of observation have shown that even after many decades some subordinate males still retained an immature appearance. This pseudoimmaturity may be strategic; appearing to be immature and nonthreatening in the presence of an intolerant dominant male may allow the small males to gain access to females. We know from taped playback experiments conducted by John Mitani that the *faux* immatures respond to the loud calls of adult resident males by moving away rapidly (for an orangutan).[22] Once they have located fertile females that belong to the resident male, these small floater males sometimes engage in a heinous reproductive tactic. They attempt to force themselves on the unwilling female, in what orangutan biologist John Mackinnon called rape. Although adult males also occasionally resort to coercive sex, forced copulation appears to be a mating strategy used mainly by immatures or transients, although the perpetrators are not necessarily successful at mating with the adult females they coerce.

This social system of territorial lone females, possessive resident adult males, and smaller transient males has defied full description for decades. What is the basic social unit of orangutan society? There may be no social unit beyond a single female and her young, surrounded by possessive but promiscuous floating males. However, there may also be a larger meta-unit that resembles an attenuated version of a chimpanzee community, in which individual females have a relationship with particular males that inhabit the same forest area. It will take further decades of field research to fully expose this larger network of relationships if they do exist.

THESE four great apes seem at first glance to have very little in common: one solitary, one harem-style polygynous, one living in communities glued by related males, and the other in communities in which female bonds may provide the glue. But underlying this variety is a deep pattern, first pointed out by Richard Wrangham in two influential papers published twenty years ago. In each of these species, as in the societies of nearly all primate species, females are a driving force. Females need particular resources in life, foremost among them food for their own reproductive health and for their gestating or maturing offspring. By living among female kin, the harmful effects of food competition are mitigated, since it is better to share key resources with close relatives than with nonrelatives. Female orangutans and chimpanzees have opted to forage for these resources on their own, probably because their ripe fruit diet is best suited to individuals or small parties eating small patches of food that could not feed a larger group. Hence, fragmenting into small subgroups as chimpanzees, bonobos, and perhaps lowland gorilla do, or just foraging alone as orangutans do, is an adaptation to feeding by females. But if females come and go as they please, how can a male hope to monopolize them for mating? Individual male orangutans try to monopolize several females for mating, but they cannot have exclusive sexual access because of their slowness and the large size of female territories. A silverback gorilla attempts to keep control over a group of females, many of whom have little to do with one another. These females transfer out of his group and into the groups of rival males. One lone male does not seem to be able to control the behavior of several females very effectively.

Male chimpanzees take a different approach; they form coalitions, often of relatives, that attempt jointly to control females. And to some extent this works, because in chimpanzee society females are not particularly supportive of one another in times of trouble, including sexual coercion. In bonobo society, however, females have power and they are able to broker relationships with males to some extent. The conventional wisdom about bonobos has been that females are dominant to males; as we saw earlier, this is true only in a feeding context and even then the male seems to be strategically deferring to females. But female bonobos are better able to assert themselves over male attempts at controlling them than any of the other apes, and perhaps for this reason male coalitions are not so marked among bonobos as they are among chimpanzees. Among the four great apes, males try to control females—that is, their movements and their behavior during estrus—but females are difficult to control because they are following their own reproductive agenda. The female agenda is to forage for food alone or in small groups so as to minimize competition for food with other group members.

Far from the passive, submissive female of the old stereotypes, female great apes are active players in the mating game. They may travel miles to obtain desired matings furtively, as shown recently in the Taï chimpanzee study by Gagneux mentioned earlier. The finding of Gagneux and colleagues that more than half of the offspring had been fathered by males from outside the social group implies that females were sneaking off to find other males from other communities to father their offspring. Since male chimpanzees are highly territorial, this is no small feat, and it is a testimony to the females' desire to expand their pool of potential fathers and perhaps future male allies by mating outside their own groups.

A second unifying feature of great ape society is the tendency for females rather than males to emigrate from the group in which they were born. In most of the higher primate species, males emigrate at sexual maturity while females remain in the natal group. In a majority of primate species, the group is composed of a core of female relatives surrounded by a coterie of unrelated immigrant males. This is what gives the lie to the old stereotype of the male-centered, male-dominated monkey group. To the contrary, group life more often centers on the lives of the females, who travel where they need to in order to find food. Males, determined to obtain mating opportunities whenever possible, seek membership in female groups. In exchange for the sexual access to females that they receive from this membership, males provide some measure of security against attacks by predators. They also protect the group's females against marauding infanticidal males from outside the group. And they may also help the females gain access to and defend desired food trees.[23]

It used to be thought that this Old World monkey pattern was true for nearly all primate species. We have learned over two decades of primate study that there are many primates that do not fit this female-bonded pattern; in many species, females and males both transfer.[24] But there is a minority of species in which males are typically philopatric (living their lives in the group into which they were born) and the females are migratory. This is the ape pattern, and although it is based on only four species it does suggest that our last common ancestor with the modern apes also lived in groups composed of related males and unrelated, migratory females. When females transfer between groups, they exert their choice of mates. Thus, in the living apes, females are far from being passive recipients of male reproductive ambitions. Rather, they engage in strategies that optimize the quality and quantity of males at their disposal.

If the relationships within groups of great apes are strongly influenced by female distribution in spite of male attempts to control female behavior, what about the relationships among groups? None of the apes live peacefully as neighbors. Chimpanzee intercommunity aggression is violent, even lethal, as events at Gombe and Mahale in the 1970s revealed. Even among bonobos we saw that, although females seek out males from other communities with impunity, most encounters between male neighbors are tense and aggressive. Male orangutans defend their females against intruding males, and silverback male gorillas have been seen fighting savagely over control of a group.[25] Joseph Manson and Richard Wrangham consider this issue in comparison to the kinds of intergroup relations that most traditional human societies have. They considered females, whether human or ape, to be valuable alienable resources for which males should and do compete intensely. This competition places male cooperation at a premium, since each male benefits by being part of a strong alliance. They conclude that this set of benefits leads to males (and men) tending to remain in their natal area and obtaining females (women) from other neighboring groups. Since males who remain in the home group tend to be related to one another, the incentive to compete rather than cooperate may be lessened.[26]

GREAT APE PREDATORY BEHAVIOR

Chimpanzees are alone among the four species of great apes in being frequent predators. We have no reason to believe that the digestive systems of the other apes could not process meat protein. Yet it is not consumed by gorillas at all, and by orangutans and bonobos only rarely. Meat, bone, and viscera surely have the same nutritional value to a bonobo as they do to a chimpanzee, so why don't bonobos hunt avidly? We used to think that gorillas were obligatory leaf eaters and so perhaps lacked the digestive capabilities to process meat, but the studies in West and Central Africa have shown them to eat a fairly chimpanzee-like diet in much of their range. There is no record of any gorilla consuming the meat of other mammals, though insects and small invertebrates are consumed readily. Surely bonobos and gorillas ought to make use of such a valuable resource whenever possible.

Consider the different reasons that the other apes might not be avid meat eaters. First, there may be basic ecological factors: some apes live in forests in which suitable prey are not available. This is rarely the case except where humans have hunted all the smaller mammals out of existence, in which case the apes' existence itself is probably threatened. It is certainly not the reason that bonobos do not hunt very often; red colobus and other monkeys are found in the same forest inhabited by these apes. In fact, a stunning observation of the relationship between bonobos and potential monkey prey was made by Jorge Sabater Pi and his colleagues in the Lilingu forest. They witnessed bonobos capturing small guenon monkeys, but rather than eating them the bonobos used them as playthings. After catching a monkey, rather than apply a killing bite and consume it as a chimpanzee would, the bonobos would carry the monkeys for hours and played with them as though they were dolls.[27] Besides monkeys, bonobo habitats in which human poaching has not been intense also contain small duiker antelope, a favorite food for chimpanzees, plus a variety of small mammals. It does not appear that bonobos fail to hunt due to lack of potential prey.

Chimpanzees make good use of their tree-climbing abilities, their powerful arms, and their large canine teeth when grabbing and subduing a monkey. The prey is usually a young monkey that is grasped by the hands, pinned to the branch, and bitten through the rear of the skull or neck. Bonobos certainly have the same adaptations for catching and killing prey as chimpanzees—as their use of monkeys as playthings attests—and yet rarely make use of them. But orangutans may be too slow to chase down prey: in their forest canopy, they would encounter mostly fast-moving animals such as squirrels and monkeys. This may be why orangutan meat eating is limited to the occasional squirrel found in tree cavities while foraging for fruit.[23] Certainly the gorillas' size might not allow them to maneuver quickly enough to catch an antelope, but the circumstances under which prey are usually met suggest otherwise. If a duiker crosses a trail in front of a gorilla, the ape need only make a headlong rush to capture its quarry. Duiker have been seen crossing a gorilla's path, but the gorilla's reaction is one of disinterest.

A third possibility is that the other apes do not hunt due to energetic constraints. The calories that a mountain gorilla might have to expend to pursue and catch a duiker might negate the caloric value of the meat. I don't find this a very plausible explanation for their failure to hunt because chimpanzees still do so. My work at Gombe has shown that the energetic balance involved in hunting rarely tips in favor of a nutritional motive. Most members of the hunting party receive very little meat for their effort, and the number of chimp-hours expended on the hunt plus the long begging and sharing session that follows it can be enormously costly relative to the quantity of meat that is usually available.[29] The most typical catch is a one-kilogram baby monkey, divided among up to twenty hunters. So chimpanzees engage in an energy-expensive behavior, and most fail to recoup their caloric investment. Granted, the nutrient value of meat is different from its pure caloric value; perhaps the value of some saturated fat and animal protein outweighs the relatively few calories that lean monkey meat provides.

These three explanations for why the other apes don't hunt are all in some way ecological, tied to the interaction between the hunter and its physical environment. But what if the motivation to hunt or not hunt is not determined by the physical environment, but by a social one? Social factors are clearly implicated in chimpanzee hunting behavior. We know that the size and composition of the foraging party determines whether its members will undertake a hunt when they meet a group of colobus monkeys. The number of hunters strongly influences whether a hunt will be successful, and the presence of particularly good hunters, such as Frodo at Gombe, can make all the difference in the hunt's outcome. These factors are not obvious in other great apes, but they may be at work nevertheless. Communal or cooperative hunting is essential to high success rates among chimpanzees, and chimpanzee societies vary in the level of cooperation that is seen. Among bonobos, fewer than a dozen hunts have been observed in two decades of research. Male bonobos do not cooperate or form coalitions to the same degree as the Gombe chimpanzees. Moreover, female bonobos are given power by males in all feeding contexts due to the strategic deference by males. If and when male bonobos make a kill, females might end up taking the meat away from them anyway. This has in fact been observed in the wild.[30] We might ask why females don't simply hunt for meat themselves if they desire it enough to take it from males. The answer is probably the same as for chimpanzees: it may be inadvisable and dangerous for a female with an infant to tackle a monkey's canine teeth or a pig's tusks. Male bonobos may receive matings from females in exchange for the meat, but if males rarely get to share in the protein bounty, then they would lack an incentive to go out and get meat in the first place. The social network of power and dominance in bonobos thus works against the use of meat and trying to catch it.

When chimpanzees were thought to be peaceful vegetarians, there was a theory that the chimpanzee digestive system was unable to break down meat, a trait that was said to distinguish them from us and also from our early meat-eating ancestors In fact, chimpanzees can digest meat as well as the human body can, although meat seems to be an acquired taste not possessed by all chimpanzee societies, nor by all chimpanzees reared in captivity.[31] It seems likely that gorillas have similar enough digestive systems that, while they are able to subsist on a diet of low-quality, fiber-rich foods, they would benefit from animal protein taken when available. But they do not. Could this be due more to the lack of cooperation between the members of a gorilla group than to any of the more traditional physiological or energetic reasons? Among higher primates, in which there are cultural traditions and a strong learned component to being a successful hunter, prey may not even be recognized as food items unless the hunters have been taught to do so. Coordination of action with other hunters is usually essential. Social predatory species such as lions, wolves, African hunting dogs, and chimpanzees all have higher success rates when they hunt together. Without the possibility of cooperation, the odds of success and therefore the worth of hunting itself may be much less. This may account for the lack of hunting in the gorilla and the orangutan, in which male cooperation does not occur.

Beyond social factors, there is culture. As I described earlier in this chapter, the cultural diversity seen among chimpanzee societies supports an emerging realization that populations across Africa at least superficially resemble very simple human societies in their variety of learned, enculturated features. The culture of hunting is one such learned tradition. Chimpanzees in an area rich with antelope might, for example, reap a bonanza of new fawns. Such could be the case in any dry, open country where there are hoofed animals as well as a chimpanzee population, such as in the open grasslands of the extreme West African portion of the chimpanzee's range. Here, we should expect meat eating because of the bounty of meat available. In other forests where there is no such abundance, the culture might never have developed, and chimpanzees might therefore not recognize antelope as potential food. We see examples of this in the failure of Taï chimpanzees to attack pigs when they are encountered, or when Gombe chimpanzees show little interest in fresh carcasses of antelope they encounter in the forest. Presumably, once the innovation of meat eating has taken place, it will spread among the members of a chimpanzee community, and into other communities with the departure of emigrating females. But cultures live and die, and it is equally likely that a learned desire for meat could be extinguished only to reemerge later. In this way some populations might hunt only particular species, or even not hunt at all, due only to the halting process of the birth and death of cultural traditions.

Finally, we must consider whether my original hypothesis is valid. Among the other apes, hunting is an unimportant part of society compared to the role it plays in chimpanzees. Since both gorillas and orangutans have been watched by human researchers for decades for long periods at close range in different types of habitat, we can safely say that gorillas do not eat meat, and orangutans do so only rarely. But nearly everything we know about bonobos comes from two study sites in the Democratic Republic of Congo, both of which have had impediments to observing hunting over the years. In Wamba, the bonobos have long been very habituated to observers and can be observed closely, but much of their behavior, especially in earlier years, has been studied in an artificial clearing in the forest where food enticements are placed. This is not a place where hunting is likely to occur. The forest at Wamba has also more recently been depleted of its wildlife by poaching, making predation an increasingly unlikely event.[32] At the other site, Lomako, the bonobos were not fully habituated to observers in the early years of research, rendering close observation of the animals over many hours impossible. We know from our experience with chimpanzees that, until the study subjects are fully approachable, predatory behavior is rarely seen, even though it may be occurring regularly; evidence of meat in feces shows this. So we must be careful about concluding that apes other than chimpanzees do not eat much meat based on the still-incomplete picture that we have of this extraordinary behavior.

The View from the Pliocene

It is clear from the study of their fossil
remains that, anatomically speaking,
the australopithecines were peculiarly
defenseless creatures.

William Le Gros Clark,
Man-Apes or Ape-Man? (1967)

When, why, and how early humans began to include meat in their diet is at the heart of our quest to
understand our origins. Diet affects every aspect of
an animal's lifestyle, and the way in which food is
acquired is influenced by a crucial set of adaptations without which the animal would not survive
another day. In the case of early humans, we can be
sure that their diet consisted mainly of plant foods,
for which they foraged all day long just as modern
apes do. We can be confident of this because our
ancestors lacked any anatomical specialization for
catching large prey, such as carnivorelike canine
teeth or powerful claws. The only other evidence of

the plant portion of the diet is the microscopic wear made by tough seeds and fibrous foods on fossilized teeth. These microwear patterns can reveal the type of plant diet eaten by long-extinct species by comparing the tooth wear patterns to those of living primates whose diet is well known. Unfortunately, we have little direct evidence of the plant food diet of early hominids. A few archaeologists, such as Jeanne Sept of Indiana University, have tried to reconstruct the likely vegetation landscape that early hominids would have relied upon, but since eating most plant foods does not require any tool use (one exception might be digging for subterranean tubers), little evidence exists of those habits.[1] The primary evidence that we have in the fossil record for the advent of meat eating is the appearance of crude stone tools, beginning in East Africa in the mid-Pliocene period about 2.5 million years ago. Before this time, if emerging humans were making and using tools, they were made of materials such as wood or bone that did not preserve in the fossil record. As to what extent the earliest hominids were eating meat without the assistance of tools, such as the chimpanzee scavenging opportunity described above, we have no archaeological record at all.

The evolutionary transition from an herbivorous primate to a more carnivorous one involves a major shift in how an animal is built as well as how it must behave. The digestive tract, armaments for prey capture, and the biting strength of the jaws are features that change when an herbivore becomes a carnivore. In human ancestors, cleverness and co-operation may have evolved to a point at which planned cooperative hunting replaced stealth and sheer attacking power. Sociality, better weapon manufacture, and larger brain size may thus be traits that arose in response to the need to find and capture prey. We assume, but do not know, that early hominids gradually became omnivores by including some meat in their diet, and that the percentage has generally increased in our more recent ancestors. Paleoanthropologists Pat Shipman and Alan Walker identify several traits that must have changed when hominids became more predator than prey.[2] These include developing ways of butchering animal products; the lengthening of the small intestine while the large intestine became shorter; an increase in leisure time as the time needed to forage for plants decreased; and attaining a low population density, since predators typically sit atop the food chain and can only live at sparse densities lest they overeat their prey supply. Some of these hypothesized shifts, such as the type of digestive tract, are untestable. Shipman and Walker suggest that the adaptive shift happened relatively recently in human history. They argue that *Homo erectus*, who appeared about 1.8 million years ago, was the first truly predatory human species. They base their assertion on the increasingly sophisticated tools associated with *Homo erectus* that may have been used for butchering prey, and on suggestive evidence that as *Homo erectus* spread its range across the Old World, it lived at low population densities in the manner of a hunting species.

How the meat was procured is as important as when it began to be a food item. Prey had to be caught and killed, or carcasses had to be scavenged. It is important for us to discover whether our own lineage arose with the help of a hunting ancestry or a scavenging way of life, since each of these modes requires a different set of behavioral adaptations. Did bands of early humans courageously attack and slaughter large and dangerous game, or did they nervously creep up to decomposing, nearly stripped carcasses to glean a few scraps of meat and fat? The route to resolving these two opposing pictures illustrates the problems and the biases involved in this field of science.

The debate about the early human diet has a long and tumultuous history. The earliest hominids almost certainly ate most of the same foods as modern apes: fruit, leaves, seeds, insects, plus some animal prey. But only the animal prey has attracted the persistent attention of anthropologists. There are some good reasons to think that humans began to include meat in their diets routinely at a very early stage of evolution, and that their ape precursors were already eating meat when they could get it. Chimpanzees routinely include small animals in their diet, and as we saw earlier, even small animals can amount to a large amount of meat when many of them are taken every year. John Yellen showed that !Kung foragers also eat large numbers of small mammals.[3] Omnivores tend to be opportunistic, grabbing a small animal or scrap of meat whenever it chances along.

Raymond Dart, whose discovery and description of the bones of the Taung child in South Africa in the 1920s was the earliest evidence of a primitive hominid, believed that the first humans were small but brutal killers who roamed the savanna with weapons in hand fashioned from bones, teeth, and animal horns. Dart believed that these hominids, which he named australopithecines ("southern ape-like humans"), used these weapons not only to kill prey but also each other. According to Dart, the australopithecines were "confirmed killers . . . that seized living quarries by violence, battered them to death, tore apart their broken bodies, dismembered them limb from limb, slaking their ravenous thirst with the hot blood of victims and greedily devouring livid writhing flesh."[4] The writer Robert Ardrey picked up on this idea and ran with it in more vivid fashion in a series of popular books that depicted evolving hominids as bloodthirsty creatures.[5]

Anthropologists ever since have been obsessed with human carnivory, despite the fact that none of the anatomy in early human skeletons reveals any adaptations for being predatory or carnivorous. The view that our humanity was molded by a predatory, warring life was forever changed when some human fossil sites showed the reverse: that emerging humans were in fact being hunted by carnivores. South African anthropologist C. K. Brain, in a landmark study, showed that the damage on australopithecine skeletons thought to have been made by other hominids was in fact caused by predators such as leopards that were feeding on these creatures.[6] The earliest hominids were thus more likely prey than mighty predator.

The activities of early humans were also easily confused with natural processes that produced similar fossil deposits, and many archaeological sites were restudied and reinterpreted during the 1970s. Archaeologist Lewis Binford and others suggested that humans were not capable of efficient, coordinated hunting of large animals until the late Pleistocene, erasing 2 million years of hunting history presumed to have occurred at that time.[7] Accumulations of animal bones and human remains that had long been seen as the product of hunters and their prey were reinterpreted as natural depositions due to stream currents and geological forces that threw a misleading puzzle together for investigators eons later.

The study of natural processes of death, decay, and deposition of extinct creatures, and the way in which they may influence the fossil record, is called taphonomy. Taphonomy began as "interesting cautionary tales"[8] but is today a major discipline within archaeology. We now recognize that patterns in the ancient human fossil record that were thought to be human-created are actually due to nonhuman causes. For example, the pattern in which we find remains of animals thought to be food for emerging hominids is an important indicator of human activities. Some parts of a skeleton—the ribs and the vertebrae—are highly transportable and also quite fragile, while parts of the skull, such as mandibles and teeth, are not. The proportion of skeletal parts in an assemblage therefore suggests whether or not a creature had been a Pliocene meal. However, this pattern turns out not to be so straightforward. Other predators, animal scavengers, and natural processes all tend to destroy the long, least-dense bones first. These bones therefore tend to be underrepresented in accumulations of fossilized bones.

Moreover, archaeologists typically try to distinguish evidence of hunting from evidence of scavenging by examining the representation of skeletal parts that are present in a fossil site. Hunting by early *Homo* is thought to be associated with a high percentage of juvenile animal skeletal parts, since the hunters preferentially chose young animals, perhaps due to the ease of killing them. Since juveniles are small and have less robust bones, those that are killed by carnivores would be largely eaten by the time the hominids would reach them. We therefore expect that accumulations of bones with low proportions of juveniles indicate scavenging.[9]

It became clear, however, that at some fossil sites humans had indeed been butchering animal carcasses, since the scratches and cuts from their stone tools were still visible on the fossilized bones of their prey. These butchers were assigned for the most part to the species *Homo habilis*, the likely manufacturers of the earliest stone tools. Anthropologists Richard Potts and Pat Shipman studied the bones of prey animals at *Homo habilis* sites. On the bones they found cut marks made by ancient sharp-edged tools as well as tooth marks made by the gnawing of long-extinct lions, hyenas, and leopards. When they examined them more closely, they saw that on some of the bones, the human-made cut marks were on top of the carnivore bites—evidence that humans were cutting flesh from the bones sometime after they had already been chewed by a predator. The implication was clear: the role of hominids was that of scavenger of the carcass of the prey, rather than hunters.[10]

This discovery represented a dramatic reversal in our thinking about the lives of early humans. To be a scavenger rather than a hunter affects every aspect of daily life. Instead of depending on the ability to chase down and kill elusive prey, a scavenger needs to find the kills made by hunters, then sneaks up to a carcass and cuts off bits or chunks before being detected. Many scavengers, such as vultures and jackals, are tolerated by larger carnivores at a kill: would early hominids have been?

Early humans may have filled a scavenger niche in African forest-grassland ecosystems between one and three million years ago. Archaeologists have adopted new approaches to understanding the role that the hominids may have played in those ecosystems. Robert Blumenschine conducted field studies of modern African carnivores to see how they turn a living zebra into a defleshed carcass, and to learn whether a creature would have been able to make a living by relying on these carcasses for subsistence. He found that early hominids would have found the carcasses of large animals killed by lions and leopards regularly, especially in woodlands located near water courses. Kills would have been quickly processed by the hunters, but sufficient nutrients remained in the carcass long enough to support a

FIG. 4.1. Lightly wooded savanna, similar to the one in which some populations of the earliest hominids lived. Queen Elizabeth National Park, Uganda. (Photo by Craig Stanford)

scavenger who relied on finding dead animals for dinner. These kills would provide little muscle tissue but much bone marrow and brain matter, and these are rich sources of fat and protein provided the scavengers can break the bones to get to the bounty.[11] Blumenschine studied the process by which carnivores reduce the freshly killed body of a large ungulate to a skeleton. A consumption sequence was followed in most cases in which the consumer rapidly devoured the hindquarter flesh first, followed by the rib cage and foreleg meat, fol-

Fig. 4.2. White-backed vultures scavenge the carcass of an antelope killed the previous night. (Photo by Craig Stanford.)

lowed by the marrow within the bones, and finally it ate the contents of the head of the prey.[12]

Assuming that ancient carnivores ate their meals in a similar way, this butchering recipe can be used as a signature in the fossil record to identify hunting and scavenging. Scavengers should eat a disproportionate quantity of a kill's last body parts containing edible meat, left behind by the hunters. Tools would have been used to get at it, and both Blumenschine and archaeologist Nicholas Toth have shown that the primitive tools made by *Homo habilis* would have been efficient cutters for the remaining bits of flesh as well as served as hammers to crack open ribs and long bones to get at the nutrients locked inside. Blumenschine considers it unnecessary and unlikely that these early hominids would have searched for meat in any other way—scavenging was a reliable enough way to obtain large quantities of good nutrition, making hunting unnecessary. Scavenging may have been a dietary niche into which early hominids shoehorned themselves. The top predators in some East African habitats in the late Pliocene period 2 million years ago were saber-tooth cats, powerful predators that could kill animals with much more meat than they themselves could hope to eat or store. Archaeologist Curtis Marean argues that *Homo habilis* could have capitalized on the remains of these kills to make its living.[13]

Whether carcasses would have been available enough to sustain hominids in Pliocene habitats is open to debate. Using the modern East African savannah or riverine woodland habitats as direct analogs for the places that early humans called home is fraught with errors. These early humans were unique species for which any modern analog, whether living chimpanzees or living ecosystems in which modern scavengers live, is only a flawed window onto the past. However, other studies have questioned whether African habitats could sustain a creature that subsisted on scavenged meat.[14] Even

Blumenschine's study in support of the scavenging hypothesis indicates that scavenging yields a lower return per carcass due to the already-defleshed nature of many carcasses. This translates into fewer opportunities for chunks of the carcass to be transported to some other place for butchering or consumption.

One of the most important hypotheses put forward related to the scavenger hypothesis is that early hominids began to use a home base. This idea was first suggested by archaeologist Glynn Isaac in the 1970s. He studied deposits of fossilized bones and stone tools on hominid landscapes, and inferred that ancient humans might have brought captured meat back to a central place for butchering it and doling it out, in much the same way that some hunting and gathering peoples do today.[15] Ancient home bases used by hominids to cache food and center their daily activities would certainly have been a benchmark in human evolution, since great apes do not return to a home base of any sort at night, merely making new nests wherever they find themselves when dusk falls. If early hominids were carrying butchered carcass meat back to a central place, many fundamental changes in their lifestyle might have accompanied the pattern. We do not know how long such home bases, if they existed, were used—for a few days, months, or centuries.

Implicit in all views of early human scavengers that I have discussed so far is that they were weaklings. Most models of the origins of early humans depict them as weaklings struggling to survive in a world that was so inhospitable that only by dint of their cleverness could they get by at all. Only a few current researchers have challenged this view. Henry Bunn of the University of Wisconsin looks at early humans and sees creatures small in stature but strong enough to have overpowered lions and hyenas to take the carnivores' kills away from them.[16] We know that wild chimpanzees will attack and harass leopards they encounter as they travel through the forest,[17] and they are not even armed with weapons when they do so. So it is not unthinkable that early humans might have outright charged at big carnivores sitting on their kills in order to drive them off and claim the meat as their own. Such piracy requires a different set of adaptations than passive scavenging, in that the ability to detect a carcass is only the first stage of obtaining it.

Bunn also points out that whereas advocates of the scavenging niche hypothesis consider hunting and scavenging as separate foraging strategies, this dichotomy is false. There are few carnivores that hunt but do not also scavenge. We view lions as majestic, noble hunters. Hyenas, meanwhile, are supposed to be cowardly, slovenly scavengers. Contrary to this image—and the Disney film *The*

Lion King notwithstanding—hyenas are actually efficient and avid hunters, while lions pirate kills away from other predators more often than they make the kill themselves. Both hyenas and lions are better characterized as hunter/scavenger carnivores than as members of only one category. Of course, each early hominid species was unique and possessed a set of adaptations that may not have occurred in any other primate, but the likelihood that any early human species scavenged extensively without hunting at all is quite small. It is easy to imagine a roving group of *Homo habilis* searching all day for fruit, and in the course of their foraging came upon either the carcass of a recently killed antelope or a very-much-alive fawn. That they would eagerly eat one but not the other seems very improbable.

Meat is such a valued resource that early humans would have eaten it at each and every opportunity. Or would they? The same ingredients that make meat such an important commodity for carnivores—a compact parcel of amino acids and calories—may also make it unpalatable, even toxic, if consumed in substantial quantity by a noncarnivore. University of Michigan archaeologist John Speth has challenged the conventional wisdom that large amounts of meat would have been eaten by early hominids. Speth points out that among modern hunter-gatherers, foods that are high in protein

are not utilized heavily during periods of food scar-
city, which is just the time when one would expect
they would be used most. This is not because meat
is unavailable during this time; instead, foragers
seem voluntarily to limit their consumption. Speth
argues that both modern people and early hominids are constrained in their meat consumption by
the toxic, even lethal, effects of consuming too
much protein.[18] Meat is a source of both protein and
calories, but only about 50 percent of one's daily
caloric intake can come from meat. Above this limit,
the liver is unable to metabolize the excess amino
acids, and the body is unable to flush itself of the
waste products of meat consumption. This can lead
to liver and kidney impairment or failure, having
lethal consequences. Furthermore, the threshold
may be lower than 50 percent for pregnant females.
Speth cites studies showing that when pregnant
women eat diets in which more than 25 percent of
calories are from meat, lower birth weights and
higher infant mortalities result.

These limits would not affect living chimpanzees—their meat intake is too low—but they could
have affected emerging hominids. During the dry
season, when the caloric intake of early hominids
must have been less than in other seasons, meat
would have become a very desired food. There is
some evidence to suggest that early hominids ate
meat mainly in drought months in East Africa.[19] But

eating too much of it may have been impossible. This, according to Speth, suggests that carcasses, whether hunted or scavenged, would have been eaten for their fat and calories rather than their protein. When chimpanzees make a kill, they typically extract the brain first, which is the single best concentration of fat in the body. This is followed by the marrow of the long bones, another rich source of fat. These bits of evidence offer intriguing insights into why emerging humans may have begun to eat meat.

All this attention we are paying to the meat of large animals ignores, however, another source of animal fat and protein that is much less glamorous but perhaps more available. Modern apes and human foragers eat insects and insect larvae whenever they can obtain them. Jane Goodall first documented chimpanzees making and using simple tools—twigs or blades of grass stripped clean and probed into the tunnels of termite mounds to coax the protective soldiers to clamp onto the tool and be extracted. Using this technique, chimpanzees routinely consume thousands of termites per day in some seasons (most often of the genus *Macrotermes*). There is a fascinating difference between the sexes in tool using. Females are much more diligent than males; William McGrew found that extended bouts of termite-fishing were done twice as often by females as by males.[20] Female chimpanzees also tend to termite-fish all year round, sometimes spending hours at a mound even though the payback in termite meals in not very high. Males, on the other hand, show little interest in termiting except at the start of the first rainy season in November in western Tanzania. At this time, the termites' reproductive caste emerges from the mounds, the earthen structures having been softened by the rains. Soldiers build tunnels to the surface but keep them blocked with soil until they are ready to be used as escape conduits by the winged, flying reproductive alates. Chimpanzees know this, and they investigate mounds to find spots where they can poke a finger into a tunnel to open it up; then they insert a fishing probe. The abundance of termites near the surface of the mounds may be so much greater at this time than during the rest of the year that male chimpanzees eagerly fish for them for hours at a time. The number of termites consumed in this way can number in the thousands at a single sitting. The sum of these tiny packages of nutrients is thought to be great by primatologists who have studied termiting. Termites and other social insects live in huge, fairly sedentary aggregations, making them sizable rewards for apes that invest the effort. Their mounds amount to enormous food patches for an industrious and patient chimpanzee. Interestingly, gorillas make less use of insects in their environment than chimpanzees, foraging for ants with their fingers at times but rarely investing hours at a time in the business.

Can the behavior of living chimpanzees inform us about the hunting-scavenging debate? We rely on the material culture present in the archaeological record to understand the behavior of extinct humans. The lesson from the study of chimpanzee hunting and tool use is, however, that the past may not always leave us clues. When chimpanzees kill an animal, they typically consume meat, bones, viscera, and hair, leaving no archaeological trace on the ground when they are done.[21] This is particularly true when the victim is a juvenile animal; it is gobbled up completely within a few minutes. Chimpanzees scavenge very rarely; in the thousands of hours of observation at Gombe, scavenging of carcasses not killed by chimpanzees has been witnessed fewer than twenty times. During the hunt of September 3, 1994, recounted earlier, the chimpanzee Freud had captured one colobus and stolen a second, discarding most of the theft without eating it. He and the rest of this party that day departed the scene after eating their five kills, leaving this carcass behind. I took the opportunity to take the carcass back to camp with me, capitalizing on a rare chance to obtain a fresh weight of a juvenile colobus (even a headless one was more than I had been able to weigh up to that time), and I also collected the many bone fragments left by the

FIG. 4.3. Part of the bone assemblage from a kill made by five male chimpanzees in Gombe National Park, Tanzania. Mandibles are most heavily represented. (Photo courtesy of Thomas Plummer)

hunters that day (fig. 4.3).[22] As we were in the middle of an expanse of thorn thicket, I crawled behind the chimpanzee party for some twenty minutes dragging the carcass before we arrived at a trail. During this time Freud and the other males glanced at me, no doubt wondering why I hadn't yet eaten it. I reached a trail, hiked to the feeding station, and placed the carcass on the ground in hopes that another adult male would come along and take or reject this opportunity to scavenge. Instead, an adult female, Gremlin, accompanied by her son Galahad, arrived first, and upon seeing the carcass at a distance, she raced to it and whisked it up a tree, where she ate it over the next hour with occasional disruptions from meat-hungry baboons.

Although the evidence is scant, females and juveniles may take meat in any form they find it, including long-dead. Males appear to be interested mainly in living prey, and perhaps also in animals they themselves have seen killed by other predators. Certainly this would match their relative abilities to obtain meat with their choosiness about the form in which they will eat it. Females are also more diligent and more skilled tool users than males. Does this combination of female interest in carcasses plus their dexterity with tools suggest that it was our female ancestors who first began to exploit animal carcasses by butchering them with primitive tools?

If we consider a protohominid in the Pliocene to be somewhat similar to a modern chimpanzee—an ape able to exploit a variety of habitats, to travel on the ground for miles at a time between feeding sites, nesting at various spots in a large home range—then certain predictions based on chimpanzees are possible. We can test these predictions against information available in the fossil record. For instance, chimpanzees do not search for their prey; instead, they encounter it opportunistically in the course of spending their days foraging for plant foods that are the bulk of their diet. Animal prey may not be actively sought because it is too energetically costly and inefficient to do so, since hunting is a risky endeavor compared to seeking the more abundant fruit and leaves. By the same logic, does this mean early hominids would not have searched for carcasses? Scavenging offers lower returns than hunting, which itself is in most cases less calorically rewarding than plant foraging. Whether early hominids ate carcasses or actively pursued live animals, or both, it is unlikely that they set out on their daily travels with the intention of doing so. Furthermore, cultural traditions may have accounted for early hominid meat-eating patterns just as they play a role in chimpanzee meat eating. East African chimpanzees relish wild pigs, which are often ignored by some West African populations. There does not seem to be anything adaptive about this, since pigs are nutritionally valuable in any forest where they are found. Carnivores tend to be divided into either prey specialists or generalists. Cheetahs on the Serengeti plains, for example, are prey specialists, eating almost entirely Thompson's gazelles.[23] They ignore freshly killed meat and the meat of other species. Since meat is meat and offers good nutrition, this specialization must have some utility to cheetahs that has been molded by natural selection—but why? Leopards, which sometimes occur in the same habitat as cheetahs, are generalists that feed on almost any source of meat that moves. Whether prey specialization is the product of natural selection or is a cultural tradition, it could account for differences in hunting and scavenging behavior between modern and extinct hominoid populations.

If early humans scavenged, they would have either bumped into carcasses accidentally—a long shot—or homed in on carcasses they detected somehow from a distance. To do this, they might have used cues. The presence of vultures wheeling overhead would be a clue to the presence of a meal waiting on the ground ahead. They also might have known when and where carnivores had been hunting lately, when birth seasons occur that would ensure a supply of helpless newborn ungulates, and where migratory routes were for animals that may die and be available as prey. All of these offer associative clues of which a clever mammal with very good spatial memory ought to have been able to make use. However, it is worth noting that it is not clear whether living apes such as chimpanzees use such associative cues. Old World monkeys apparently do not; vervet monkeys do not make a connection between the presence of a fresh carcass and the potential danger of the carnivore that killed it lurking nearby.[24]

When chimpanzees make a kill, they create a taphonomic site, and we can study predation landscape for either modern apes or ancient hominids. In places like Olduvai Gorge, deposits of bones and stone tools are found together. The interpretation—that hominids used the tools as weapons to kill prey—is fraught with problems. We can use the killing landscapes made by chimpanzees for modern-day case studies that suggest patterns our ancestors in the Pliocene may have followed. For example, by studying the kill site made by Gombe chimpanzees in September 1994, we can learn how an observed predation event leaves a record. Unfortunately, chimpanzee kills leave almost no archaeological trace, since they consume every bit of the carcass in most kills. We can, however, use some aspects of the pattern of chimpanzee kills in comparisons with existing fossil sites. Our work at Gombe shows that the parts of the carcass most frequently left after a kill is eaten are the skull and jawbones. In the kill of September 1994, for example, at least three individuals are represented in the bone assemblage, although five kills were actually made. The maximum distance between the spots where the hunters sat consuming the meat and scattering the bone fragments was about ten meters.[25]

Using what we know about the hunting ecology of wild chimpanzees, we can reconstruct the likely similarities and differences with the predatory patterns of the earliest hominids. The similarities are as follows:

1. The diet was primarily plant foods, with meat contributing a relatively small percentage to the overall diet. Prey, whether captured alive or encountered dead, were simply located opportunistically during the daily course of foraging for plant foods.

2. Meat was probably eaten seasonally when other foods were not available, and perhaps when prey birth seasons provided a bounty of fawns or infants. In the leanest season, the portion of meat in the diet may have been substantially greater than the yearly average.

3. The hunting range or territory was the home range of the group, and the frequency of hunting was related to patterns of range use. Hunting or scavenging was most frequent within a core area that was smaller than the total range. The main influence on travel patterns was the vegetation landscape that created a mosaic of plant food foraging opportunities, not the possible but uncertain locations of meat sources.

4. Hunting was done mainly by males working together. The odds of a successful hunt were directly proportional to the number of hunters involved in the pursuit.

5. After a kill was made, the prize was shared by some but not all members of the hunting party, and also with females that were present.

6. The bone marrow and the brain were the favored parts of the prey.

7. The possession and control of meat were used by males for their own selfish political and reproductive gains.

8. Prey were mostly small to medium sized—under 40 kilos (90 pounds). Nevertheless, the total amount of

prey killed and scavenged each year was great (more than 220 pounds per adult male in the social group per year, and smaller amounts for females and adolescents). Early hominids may have been major predators on small and medium-sized mammalian species in their habitat, especially if their hunting peak was concentrated in the birth season of the prey.

There are, however, at least two profound differences between the meat-eating behavior of modern chimpanzees and that of emerging humans:

1. Stone tools used for butchering prey carcasses came into use about 2.5 million years ago.
2. The size of prey increased dramatically sometime after stone tools entered the fossil record, and with larger prey came the greater importance of meat in the diet.

The list of similarities therefore applies chiefly to the earliest hominids, before the documented use of stone tools. Without having an effective means of cutting up prey, these creatures may have had to rely on small animals that were caught alive or medium-sized carcasses that were scavenged. However, even small animals—rabbits, monkeys, piglets—can add up to a very large quantity of meat if enough of them are eaten. It is also essential to remember that when we speak of "early hominids" we are really talking about a number of species, each with many populations having unique adaptations. The behavioral ecology of the earliest hominids may have been very different from that of their *Homo* descendents 1.5 million years later.

All of the foregoing applied to the pattern of meat acquisition very early in the human fossil record, from 2.5 million years ago, when stone tools first appeared. But hunting and scavenging patterns of our more recent human ancestors may tell us much about the evolution of hunting and its role in human evolution, since it provides a link with modern, traditional hunting peoples. We imagine more modern hunters to be very different—after all, they are men engaged in the pursuit of big game armed with weapons, whereas the early forms of hunting discussed in this chapter apply to apelike creatures that use only their teeth and hands.

One might expect the rise of more modern humans, with a related expansion in brain size, to be accompanied by cooperation during the hunt, greater powers of either spoken or gestural communication, and overall better hunting efficiency. *Homo erectus*, an early human species that lived in the Old World from about 1.8 million years ago to 300,000 years ago, has often been considered the first big-game hunter, capable of using fairly sophisticated weaponry to bring down animals as large as elephants. It was once an accepted, or perhaps assumed fact, that humans of the early Pleistocene era

were accomplished hunters. Perhaps the one archaeological site most persuasive on this account
was Torralba, an ancient lakeshore on the plains of
central Spain. There, *Homo erectus* was thought to
have hunted elephants. Early studies at Torralba by
F. Clark Howell documented massive numbers of
elephant bones, which were deposited in the same
locations as the remains of an ancient human culture.[26] Hand axes, tools as characteristic of *Homo
erectus* as the computer is of modern society, are
present in large numbers along with tools made of
bone. Howell noted signs of ancient fires at the site
and suggested that the early hunters intentionally
set fires to drive their huge prey into the mud at the
edge of the lake. Using weapons and game drives,
Homo erectus appeared to have been a clever and
very successful predator.

This conclusion changed only as the earliest ta-
phonomists rose to the fore. Lewis Binford and his
student Leslie Freeman took on Torralba as a place
where the evidence might be looked at from different angles, without the prior assumption that cooperative human hunting was responsible for the site.[27]
In essence, Binford both stepped aside from assumptions about human prehistory and pursued
his own theoretical model of how the natural world
works in making a fossil site. These two processes
are far more easily described than accomplished.
Speaking about the earlier interpretation of Tor-

ralba, Binford states: "This is a classic example of a post hoc accommodative argument: that is, the a priori assumption is invoked to give meaning to the recognized patterns."[28] The initial interpretation about Torralba as a site of cooperatively organized human hunting activities lasted for years, and all further investigations of the site were bound by that set of presumptions. Only later did many of the interpretations become revised: bones that were thought to have been modified as tools were actually molded by the action of water in the lake. Tools that were apparently fashioned from elephant tusks were merely tusks exhibiting natural breakage patterns. Even the degree of association between human and elephant bones was called into question. The result was that the early interpretations about Torralba were overturned. The assemblage of elephant bones appeared to be the result of natural processes rather than cooperative hunting. Thus, even when dealing with highly evolved humans known to use fire, make weapons, and hurt animals, we must be cautious about making interpretations regarding social hunting.

It appears that the emergence of a more modern human pattern of hunting of big game did not arise in the Upper Pleistocene in Europe until some 110,000 years ago. All populations known after this time are ambush hunters who probably captured game in much the same way as modern foragers like the Hadza in northern Tanzania. We infer this from the fossil record by the representation of the prey population found in assemblages of bone in archaeological digs. In one such study, archaeologist Mary Stiner found that Upper Paleolithic people in Italy were butchering and eating healthy adult animals. This pattern suggests hunting rather than scavenging, since scavengers more often rely on finding the carcasses of the old, sick, and young.[29] Eating prime adult prey habitually is unique to humans, because prime-age adults are the most difficult age class to catch and kill. Capturing a prime-age animal may also require more cooperation than catching young or old prey. In addition, prime healthy adults are likely to be the best sources of stored fat. These features make adults attractive as prey, but they are a difficult commodity to obtain. Stiner's work on the Paleolithic of Italy indicates that slightly earlier in prehistory, people living in Europe both scavenged and hunted. The archaeological record for recent human prehistory thus seems to show that our ancestors emerged from a past in which meat was gotten by whatever manner possible, into a recent past and present in which hunting for live adult prey has become the norm.

This brings us closer to the present to consider the nature of meat foraging by more recent humans in the Pleistocene. We have seen that recent human ancestors shifted away from scavenging, whatever

its part in the diet may have been, toward a pattern
of ambush hunting for larger animals. The way in
which nearly modern humans hunted may tell us
about their cognitive abilities as well as their ecological role as predators. The Neandertals were a
form of early humans who lived in the Old World
from about 100,000 to 35,000 years ago. Some archaeologists and physical anthropologists are skeptical that Neandertals were organized hunters capable of coordinating a group hunt. Some of this
thinking is subject to the same sort of logical fallacy
that causes paleoanthropologists to dismiss meat
eating by great apes. Erik Trinkhaus of the University of New Mexico, for example, sees in the robust
Neandertal anatomy evidence of a lack of efficiency
of travel and hypothesizes that Neandertals were
inefficient, disorganized hunters, roaming the landscape in an often-futile search for animal prey.[x] He
interprets this from anatomical aspects of Neandertal lower limbs, which suggest irregular, inefficient
locomotion. Trinkhaus sees Neandertal brawn as an
evolutionary compensation for a lack of intelligence. But chimpanzees, monkeys, and other mammals manage to navigate their habitats efficiently in
search of food with spatial memory and intellect far
below that of Neandertals or any other primitive
human. Whatever the reason was for the Neandertals' adaptation to great strength and endurance, it

is highly unlikely that it reflects a lack of intelligence or planning in hunting on their part.

The fossil record is not entirely clear on when humans made a transition from pursuit hunting to the sort of ambush hunting that requires a greater degree of planning and may indicate higher cognitive development. This transition may have happened with the advent of anatomically modern people about 100,000 or more years ago. Guy Straus of the University of New Mexico sees a quantum leap in the degree of planning of hunts starting about 20,000 years ago in Europe and southern Africa. Fully modern *Homo sapiens* in Europe ate not only meat of ungulate mammals but also products of the sea and lakes. Their mainstay was Europe's big mammals of the time: bison, reindeer, red deer, wild horse, and ibex. These are large, herding, potentially dangerous animals, whose killing would require a major cooperative effort. Reindeer in particular were a favored prey on the plains of northern Europe. Hunters appear to have targeted reindeer at northern climes, while those in warmer, southern climates exploited a wide range of animals and plants in their environment.[31] In either place, we know that these hunters were armed with sophisticated weapons, spears, darts, perhaps bows and arrows, harpoons, and other sorts of projectiles. This gave people the ability to bring down creatures at which they could only stare longingly earlier in prehistory. Humans living 20,000 years ago had also learned to plan their hunts in advance, and to drive game through narrow bottlenecks in the landscape where they could be more effectively ambushed. They may also have known the annual migratory cycles of the hoofed game and followed them, or planned their hunting calendar around the much-anticipated movements of game through their area. Hunting by more modern men thus becomes routine, strategic, and well coordinated rather than opportunistic. To understand the importance of this shift, we must consider the meat-eating behavior of modern foraging people.

5

The Hunting People

> When you come home empty-handed, you sleep and you say to yourself, "Oh, what have I done? What's the matter that I haven't killed?" Then the next morning you get up and without a word you go out and hunt again.
>
> A /Xai/xai San hunter,
> quoted in R. Lee,
> *The !Kung San* (1979)

In a few scattered pockets of the world, there are people who continue to live in intimate contact with the land, using only a relatively simple technology combined with their considerable environmental wisdom and intelligence. These societies, which we call hunter-gatherers or foragers, are nearly gone; the total world population of foraging societies today numbers only in the tens of thousands. It includes people in the Canadian arctic, the Amazon basin, and the savannas and rain forests of Africa. These people have been widely used as portraits of what our ancestors may have been like. They teach us about the interaction between human behavior and environment, and in doing so inform us about the range of possibilities of social adaptations found in their, and our, ancestors. They gather plant foods, honey, insects, and many varieties of small animals from their habitat, and they also hunt (table 5.1). In nearly all foraging societies that have been studied, men do most of the hunting and women do most of the gathering. They are, in a sense, humans living in their natural habitat.

Before describing the hunting and meat-sharing patterns of modern foragers, there is an important question to consider. Is the depiction of modern people who live with simple technologies as "primitive" tainted by a flawed romanticism, and perhaps even racism? While foraging societies possess a simple array of technologies, it would be a profound mistake to consider them "primitive" societies. Rather, these are people who subsist upon the land, without connection to a cash economy.

It has become fashionable for anthropologists to criticize the use of foraging people as models of human evolution, citing all the ways in which they are not appropriate illustrators of early human behavior. If we compare modern hunter-gatherers to a likely portrait of early humans living one to three million years ago, major differences exist that should serve as caveats to drawing evolutionary

TABLE 5.1. Some tropical and subtropical foraging societies and their diets.

Group	Region	Percentage of Diet from Hunting Animals (by estimated calories)	Percentage of Diet from Gathering Plants and Other Foods
Aché	Amazon basin forest	44%	55
Piro	Amazon rain forest	14 (includes fishing)	86
Efe	Central African rain forest	8.5	91.5
Dobe San	Southwest African desert	40 (seasonally 20–90%)	60
(G/wi)	Southwest African desert	15	85
Hadza	East African savanna	35	65
Siriono	Amazon basin forest	25	75
Walbiri	Australian desert	30	70

SOURCES: Aché—Kaplan and Hill 1985a; Piro—Alvard 1993; Efe—Bailey and Peacock 1988; Dobe San—Lee 1979; G/wi—cited in Kelly 1995; Hadza—Woodburn 1968; Siriono—Holmberg 1950; Walbiri—Meggitt 1962.

conclusions from modern hunter-gatherer behavior. First, early hominids possessed a different anatomy from modern people, which no doubt influenced every aspect of its behavior. Earliest humans were technologically far simpler than modern foragers, and were probably closer to what a wild chimpanzee does with tools today. Although we like to think that hunter-gatherers live in a primeval environment, that is almost certainly untrue. For hundreds of years most foraging societies have been in contact with other, nonforaging people. The !Kung, for example, were long considered to be ideal models for the roots of our own behavior. We now know that they had trading relations with the outside world for hundreds of years before their contact with European and American anthropologists.[1] The Efe pygmies of the Ituri forest of eastern Congo appear to be the ultimate forest people. But field studies by Robert Bailey and Nadine Peacock have shown that the Efe are very much dependent on another group of people, the Lese, for trading relations with villages outside the rain forest. Without this cultural symbiosis it is not clear that the Efe could survive in the rain forest.[2] Perhaps rather than presume similarities between ancient and modern people, we should focus mainly on testing likely differences.

Many anthropologists therefore assume that we should not use modern-day hunter-gatherers as ex-

amples of the range of early human social and ecological adaptations. This reluctance is entirely un-
warranted. There is no other large, highly social,
tool-using biped except ourselves. This is what
Homo habilis was 2.5 million years ago. These parallels alone make modern foraging people the best
and the only living examples of what humans do
when confronted with the forces of natural selection through the rigors of the natural world. Traditionally, anthropologists have tried to understand
the world of the forager through cultural explanations, the set of learned traditions that compose the
fabric of all societies. Humans are of course biological organisms, and our capacity for culture is as
fundamental a part of our biology as our bipedal
posture and oversized brain. Modern behavioral
ecologists have used the biological roots of culture
to ask whether the rules that govern the behavior of
other primates may not also shape the behavior of
humans.[3] For the foraging people, food-getting itself is a critical goal that affects each individual's
survival and reproduction. So we can ask whether
foraging for plant food and for meat, plus exhibiting social behaviors that affect successful foraging,
are subject to natural selection as in all other primates. Energy and time spent searching for and
capturing animals must be offset by the caloric and
nutrient value of the animals that are caught. This
bottom line provides a currency with which we can

try to understand the decisions that foragers make each day about which foods to search for and which to ignore.

The sun rises over the rift escarpment in Gombe National Park in Tanzania, finding a party of chimpanzees already out of their night nests and heading off for a day of travel and feeding. By the time darkness comes again, these foragers will have traveled several miles in search of food. Their route is circuitous, winding around valley bottoms before beelining up precipitous slopes, cresting ridge tops and then plunging into the next valley. They stop at fruit-laden trees and appear to know exactly where these arboreal supermarkets are located. Between such pit stops, they pass other food sources on which they may nibble until ready to go on. Several times during this particular day they encounter meat: two colobus groups during the morning and a baby bushbuck hidden in tall grass just as the sun is setting. These prey were not sought out; they were encountered only because the party's compass point took them by chance into the path of a potential meal. The colobus groups are ignored, but the chimpanzees do not pass up the easy bounty of meat from the fawn, which has been left in a grassy thicket by its mother. The hunters whisk the baby antelope from its hiding place and carry it off with screams of excitement, anticipating to devour it in a nearby tree. As night falls, the carcass is nearly gone, and a few of the hunters take bits of skin and flesh to their nests, where they will con-

tinue to nibble on it during the night. The day's energy-balance tally sheet: 3 miles traveled, nine long feeding stops at fruit trees, and numerous other short pauses, plus the meat of the fawn.

* * *

Another forest, another higher primate. Near the border between Brazil and Paraguay, a group of Aché men are spending their day on a foraging sortie. As dawn arrives, the men in camp awake and begin sharpening arrows and rehoning their weapons for a day of hunting. As they discuss and disagree on the best direction to take through the forest in hopes of finding prey, capuchin monkey calls are heard to the south. Two men leap up and head off in their direction, giving calls themselves. By imitating the infant distress calls of the monkeys, they hope to lure the group in, or at least keep it from moving away. A few minutes later, the rest of the men in camp suddenly get up and leave, walking quickly in single file in the direction of the capuchins. The women stay behind; they will pack up the camp and follow the men at some distance throughout the day. Shortly after departing, the men split up. They proceed in ones and twos in the same general direction, repeatedly spreading out and coalescing. They may locate beehives, valued plant foods, and other small animals that can be collected while the foray continues. Occasionally a lone hunter will flush an armadillo or a wild pig. He then calls his fellow hunters, who converge on the spot and dispatch the prey, sometimes tracking a wounded animal for great distances. Finally, several hours after they have left camp, the foraging party meets a capuchin group. The five-kilogram monkeys are common in the forests of the Aché, and are hunted nearly every day. The capuchins flee at the approach of the humans; the hunters pursue on the ground below, stopping repeatedly as they look for a clear bow shot up into the trees. All other hunters in the vicinity are called in, and some of the men climb into the trees and take aim with their bows. An arrow is shot, a monkey is hit and falls from the tree canopy, thudding into the earth where it lies injured but not dead. A hunter rushes over and grabs the animal, throttling it by the neck until it stops moving. In this way ten monkeys are killed. As dusk comes the hunters have met the women and children at a new camp, the meat is butchered by men and women together and cooked over an open fire. The meat acquired that day is then doled out liberally to the other families in the foraging party.[4]

There are profound differences but also striking similarities between the foraging ecology of humans and apes. Of all the differences in foraging trips by human and nonhuman hunters as described above, one of them should be most obvious. While the prey that provided meat for the hunters is equally relished by both humans and chimpanzees, only the humans actively searched for their prey. For chimpanzees, the hundreds of prey animals that are eaten each year are essentially a bonus of protein and fat that comes to foragers in search of plant foods. For hunter-gatherers, foraging trips have the explicit intention of finding both meat and other foods. I have seen very few encounters between chimpanzees and their prey in which the chimpanzees changed their travel direction or in any way indicated an intention to find prey rather than simply bump into it while looking for other foods. This seems strange given the obvious desire that chimpanzees have for meat. Why would human hunter-gatherers search for sources of meat while chimpanzees do not?

The answer may lie in the availability of meat in a tropical forest, and how humans compared to chimpanzees travel in order to locate their food. Meat, whether it is living animals or the carcasses of dead ones, is almost certainly more widely and unevenly scattered on the landscape than plant foods. Even a rare and sought-after species of fruit will be found more easily and predictably, simply because the fruit tree is stationary and usually bears its bounty at predictable times. Prey animals move around, and even dead ones lie out of sight while their bodies decompose into inedibility. Meat sources must be located and reached fairly reliably and quickly. Great apes certainly have the cognitive capabilities to remember where they have recently seen good food of any kind and know when and where to look for desired plant foods. But getting

there is a different story. Chimpanzees travel quadru-
pedally on the forest floor; very little travel is done
overhead in the trees. On the ground, chimpanzees
knuckle-walk, a mode of travel that is inefficient compared to the gait of other large-bodied mammals.
Even though chimpanzees may walk several miles
per day in the wild, they are generally traveling to
known locations of fruit trees and don't knuckle-walk
aimlessly around the landscape. Chimpanzees are no
well suited for the sort of low-reward, long-distance
searching that finding meat would entail.

Human foragers possess the bipedal posture and
gait, enabling us to walk slowly but efficiently for
many miles. Many human foragers travel farther on
a foraging day than chimpanzee parties ever travel.
The origins of our bipedal posture are not well understood, but walking upright has made all the
difference in many aspects of our behavior, and
meat-eating may be one of these. When we speak of
hunting effort by chimpanzees, we are only considering what they do after happening upon a prey
animal. For humans, the equation is entirely different because the energy invested in foraging for
meat can involve a large amount of searching effort
beforehand, though other foods may be located
during the search as well. This suggests that when
early humans became bipedal they began to search
for meat actively, because only then did it become
energetically feasible to do so.

Humans forage with an entirely different plan in

mind than apes, made possible by their ability to combine dietary items that offer a high return for foraging effort along with low-return fallback plant foods. Humans travel long distances efficiently to find food, then cooperate to increase individual return rates, and also employ weapons as killing tools. What do chimpanzees do when they hunt that human foragers do not? Foremost, chimpanzees make use of the forest canopy and their wonderfully adapted-for-climbing anatomies. They can mount an attack on monkeys high above the forest floor that would be out of the reach of human hunters armed with arrows. Chimpanzees also possess their own weaponry: long canine teeth that can puncture the skull of a small monkey, pig, or antelope. Although chimpanzees use tools effectively for eating insects and also incorporate branches and rocks into their charging displays, they virtually never use these tools to help kill the animals they are pursuing. Humans, on the other hand, routinely use bows and arrows, blowguns, nets, clubs, and spears to facilitate making the kill.[6] Human hunters routinely kill animals much larger than any that a chimpanzee would contemplate tackling, and weapons may make this feasible. The largest prey ever recorded for a Gombe chimpanzee was a large bushbuck fawn of about 20 kilos, and most chimpanzee prey consists of small monkeys that weigh less than 5 kilos (11 pounds). In addition, humans

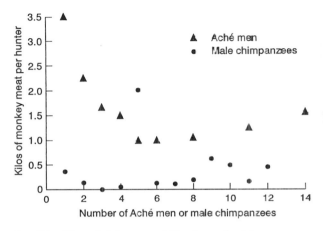

Fɪɢ. 5.1. When Aché men and Gombe male chimpanzees hunt monkeys, the effect of the number of hunters on the success rate is similar: hunters get more meat in the smallest and largest parties. Note that Aché men catch larger quantities of meat overall. Success is measured as the number of kilos of meat caught per hunter once the prey is encountered.

use weapons to kill prey that are high in the forest canopy above them. Without a weapon, much of the meat protein in a tropical forest would be unavailable to a hunter-gatherer. Chimpanzees, while ably equipped to kill prey and possessing a great relish to do so, do not rely on meat like even the most meat-impoverished human societies do. Although hunter-gatherers hunt alone or in groups, cooperation among the hunters can increase hunting success dramatically (fig. 5.1).

Cultural differences between chimpanzee populations pale in comparison to the rich variety of
hunting techniques and meat-eating cultures that
exist in traditional human societies. Among the
!Kung San of the Kalahari desert in southwestern
Africa, hunting can involve either mobile, long-distance searches for game or stationary rope snares
that are set to trap animals. Snaring brings lower
returns than searching, and is not favored by men
who are old enough and strong enough to depart
camp on hunting treks. Men are usually the hunters
but women often accompany the hunting party. In
Richard Lee's study of the !Kung San, the mobile
hunter carries all manner of equipment on the trek
that prepares him for any eventuality. He relies on
his considerable ecological wisdom of the environment as well as spiritual help from his gods. He carries a hunting tool kit and is prepared to locate and
track a game animal for many miles over many
days if necessary.[7] To the !Kung, "big game" means
animals that weigh more than 100 kilos, which is far
larger than even the largest prey taken by apes.
Prey as large as this is almost always killed by men
hunting in groups. These hunters coordinate their
actions through hand signals, and while one hunter
stalks the animal at close range, companions block
possible escape routes. The goal of the first hunter is
at least to wound the prey while chasing it in the
second hunter's direction. Prey are usually not

killed outright; they are wounded and tracked for
long distances, and a large percentage of them are
never found.

The Hadza of northern Tanzania are a well-studied group of foraging people who live in the arid
grasslands of the great Rift Valley of East Africa, the
crucible in which the ingredients for ancient humanity were brought together. This, plus their lifestyle, has made them prime research subjects for
anthropologists interested in how earlier forms of
humans living in the same place in a similar habitat
would have made their living. Anthropologists
Kristen Hawkes and James O'Connell and archaeologist Henry Bunn have conducted research on
how the Hadza obtain meat and what they do with
it. Hadza men adopt different hunting strategies depending on the season. In the wet season, Hadza
men go off on foraging excursions and encounter
the whole range of animals that share their habitat,
stalking their prey with poisoned arrows. In the dry
season, they adopt a different tactic; by sitting in
wait in blinds near water holes, hunters are able to
capture prey that come to drink. In addition to these
two methods, Hadza men also scavenge the carcasses of game animals. When other predators such
as lions or hyenas make a kill, the Hadza watch for
the signal flights of vultures over the area and also
visit areas where they believe carnivores have been
hunting. Upon finding a carcass, the Hadza employ

what may be a time-honored method of obtaining meat; they drive off the rightful possessor through whatever means available and take the carcass.[8] This piracy is far different from passive scavenging, in which they simply happen upon carcasses while looking for other foods. Scavenging provides a meat supplement to the prey animals that the Hadza can kill, but it does not provide enough meat to be a primary mode of existence for them.

It has long been believed in some quarters that because of their hunting techniques, indigenous hunting people are the world's original conservationists. This view supposes that hunter-gatherers are aware that the land will sustain them so long as they respect the balance of the ecosystem, and therefore conserve their resources in ways that Western industrialized societies have long abandoned. Unfortunately, this view of the ecologically "noble savage" is far from accurate. In the western part of the Amazon basin, the Piro Indians of Peru are avid hunters who also fish and cultivate crops. Michael Alvard investigated the possibility that these people choose their prey in a way that would prevent the long-term depletion of prey populations, suggesting a conservation-oriented strategy. He found the opposite; Piro decisions about what species to hunt in no way seem to reflect a conservation ethic. Larger animals such as monkeys and peccaries, which provided good return rates of meat but whose reproductive rates were slow enough to endanger their populations, were hunted intensively nevertheless. Furthermore, a conservationist should eat immature animals preferentially, thereby saving the breeding adults to provide for future generations. But the Piro killed adults and immatures alike. For the Piro, then, short-term gains prevailed over long-term harvesting strategies that would conserve prey populations.[3]

Wild chimpanzees eat a diet that contains a percentage of meat so small that we cannot be sure that it plays any role in their nutritional well-being. But for human foragers, it seems clear that meat is both highly desired and highly beneficial nutritionally. Why successful hunters even bother to gather, and why they would try to include certain resources in their diets, is not clear. It is therefore important to understand why hunters hunt in the first place.

Foraging people do not hunt for sport; they kill only to obtain meat for direct consumption, to feed their kin or others in their community. The nutritional rationale for hunting seems obvious, but we really do not know what nutrients are most sought-after in the meat that traditional foragers hunt. Protein is a possibility, and meat is a denser parcel of protein than could be gotten from any plant food. Protein seems unlikely to be highly valued as a source of energy; as we saw in chapter 4, the body must crank up its metabolic rate a notch just to

make use of protein.[10] Fat is another prime candi-
date as the most valued resource; it seems clear that
chimpanzees value fat above all else in the carcasses of their prey.[11] We infer this from their preference for the brain and the bone marrow, two of the
most fat-rich body parts. Fat, however, may be less
desirable as a source of energy than other foods.
Every athlete knows that carbohydrate-rich foods
such as fruit yield energy almost immediately,
whereas fatty or high-protein foods require a pro-
longed wait. Humans might also value trace min-
erals that cannot be gotten elsewhere. Perhaps all
are seeking calories from meat.

The Aché people of eastern Paraguay, whose foraging trip I described at the outset of this discussion, are a well-known New World people who forage for a living and consume large quantities of
meat. Anthropologists Kim Hill and Kristen Hawkes
have studied the Aché since the late 1970s and have
documented the details of their foraging strategies.[12]
Using an evolutionary framework to understand
why the Aché make foraging decisions that include
some foods and not others, Hill and Hawkes found
that the Aché hewed closely to a balance sheet of
energy and time expended to calories obtained.
Meat is a fine source of nutrients, but not a reliable
source of caloric energy compared to plant foods
that can be gathered more easily. The Aché are un-
usually carnivorous for a foraging population; their

caloric intake from meat is higher than nearly any other foraging society that has been studied.[13] Hawkes and her colleagues ask, "Since the Aché do so well hunting, why do they gather?" For most foraging people, plant foods may represent a hedge against failure to obtain meat. Meat is a highly desired food resource, but one that carries a high risk of failure and a low return on time and energy spent in its pursuit. Plant foods can be gathered from the forest, often by women while men are hunting, at little risk of failure and with high returns on the investment of time and energy. For the Aché, plant foods are not simply a low risk/high return insurance resource against failure to get meat; foraging for plant foods (and insects) is weighed into foraging decisions because these are foods that have lower costs in time and energy relative to searching for difficult-to-acquire animal foods. When foraging is seen as a set of economic cost-benefit analyses, it becomes clear that foragers should hunt sometimes—such as when the prey is abundant and easily caught so that return rates are high—and gather at other times.

In order to explain why the Aché hunt, Kristen Hawkes has suggested that we should consider what the Aché do with meat once they have captured it. When hunters come back to camp after a successful hunt, parcels of the catch are doled out liberally to families. The captor of the meat does not necessarily end up controlling its distribution, and his attempts to control it are not tolerated by the group. More of the largesse ends up in the hands of other families than in the captor's own household. This apparent altruism may be self-serving, as it hedges against the day when the hunter himself fails to capture meat but receives a reward anyway. Sharing may be a strategic way to lower the wide variation in the odds of catching one's own meat every day.[14] Hillard Kaplan and Kim Hill showed that among the Aché, some foods are shared more often than others, especially the large game animals. Big game is perfectly suited to sharing among hunting families, being too large for one family to use all at once and unpredictable in their occurrence. Nutritionally, Kaplan and Hill found that the advantage of receiving shared meat from unpredictable game was greater than the cost involved in sharing from one's own kill.[15] Paradoxically, even though some hunters contribute huge quantities of meat to this public good and others very little, there was no correlation between bringing home a great prize one day and receiving an equivalent or larger share in return at other times. In other words, if this is a system of cultural reciprocal altruism to reduce the risk to everyone of going hungry, it still does not explain why big game is shared so widely.

There is a paradox here: Why should a man go out hunting at all if his success does not increase the bounty he will eventually receive from other hunters? Kaplan and Hill consider four possible explanations for Aché food-sharing patterns.[15] First, we should expect meat sharing to follow lines of kinship based on evolutionary principles; individuals should behave altruistically mainly toward relatives with whom they share genes. The kin-selected basis for sharing means that the sharer is not being entirely altruistic in giving up most of his catch. He may be aiding his kin in feeding his nephews and nieces, who in return pass on genes that are shared by the hunter to the next generation. However, Kaplan and Hill found that Aché did not receive more meat from hunters who were kin than those that were not relatives. Second, the anthropologists asked whether shared foods are most often those whose capture requires a team of hunters rather than one man. Peccaries and monkeys, for instance, are rarely caught by lone hunters because of the need for cooperation in bagging these prey. But the percentage of food that is shared outside the hunter's family is exactly the same whether the prey was caught by one man alone or by many men working together.

The third hypothesis for sharing is based on the almost universal observation that meat sharing among hunting and gathering peoples occurs because the possessor of a prime piece of meat is relentlessly begged and harangued into giving up some of his catch. Kaplan and Hill hypothesize that this strategy of obtaining meat works through a model known as "tolerated theft," a term coined by Hadza researcher Nicholas Blurton-Jones. Hunters may surrender portions of their catch not out of altruism but because, under persistent pressure from the begger, the cost to the captor of continuing to defend the entire prize is greater than the cost of simply giving a few scraps away in order to be left alone.[17] This model works best when the hunter acquires a large animal that he cannot fully use, anyway. However, the Aché hunters routinely give up so much of their catch to others that there is rarely a sharing-under-pressure scenario in the aftermath of the hunt. Finally, Kaplan and Hill ask whether sharing could be accounted for by reciprocal altruism. The occasional capture of large packages of meat protein mean that giving some away today is a good strategy to get some in return at a later date when you have failed to get food. However, the better hunters still end up giving away much more food than they can ever expect to receive. Unless the tit-for-tat system of reciprocity is balanced over a longer period than the researchers measured, it does not provide a satisfactory explanation for sharing, either.

Hawkes takes an alternative approach to explain the paradox of Aché sharing patterns. She questions whether the purpose of the hunt is entirely nutritional, since the captor seldom controls the distribution of the carcass and therefore ends up with only a minor portion of it. Hawkes points out that by hunting, men are targeting resources that are shared widely and are therefore available for use as politically strategic tools. Males also have a different set of Darwinian priorities than females. While females can create better reproductive opportunities for themselves primarily by bettering their own and their offspring's food intake, males depend on having access to as many females as possible for their own reproductive benefits. This follows from the observation that females have a limited potential for reproduction and so tend to be sought after by males, rather than vice versa. Males, on the other hand, are limited reproductively only by the number of mates they can acquire and often need to invest little or no time and energy in raising the ensuing offspring. Among the Aché, men show a strong preference for obtaining the foods that are the most widely shared, suggesting that they have something to gain from its acquisition.[18]

Hawkes reasons that Aché men are showing off by sharing a bonanza of meat. These men receive many benefits from providing a public good from which few are excluded. Among these benefits may be a man's opportunity to enhance the number of children he produces. Kaplan and Hill found that Aché women prefer the best hunters as their partners in extramarital affairs. Women consider good hunters to be more desirable as sexual partners, if not necessarily as husbands.[19] The sharing of big game meat by Aché men may therefore fall into the category of a mating tactic rather than family provisioning. Kaplan and Hill are careful to point out that being a good hunter does not necessarily make a man more successful in leaving many descendants. But even if affairs are only a cultural effect of bringing home bonanzas of meat, there are likely evolutionary effects that follow from this behavior.

Sharing and Egalitarianism

The division of spoils lies at the heart of the hunt for many foraging people. The sharing of hunter-gatherers fits a pattern of giving to those around you who in other societies might be considered to have higher status. This generosity has often been linked to the egalitarian nature of forager society, in which modesty and consensus striving are valued above status seeking:

> "Say that a man has been hunting. He must not come home and announce like a braggart, 'I have killed a big one in the bush!' He must first sit down in silence until I or someone else comes up to his fire and asks, 'What did you see today?' He replies quietly, 'Ah, I'm no good for hunting. I saw nothing at all . . . maybe just a tiny one.' Then I smile to myself because I now know he has killed something big."[20]

Among the !Kung and the Hadza, even the best
hunter must be modest. Humility is a strong cultural tradition in nearly all hunter-gatherer societies. This is an extension of the egalitarian nature
of these cultures. Attempts to use one's hunting
prowess as an entrée to greater ambitions within
the society are usually met with stern opposition,
ridicule, and attempts to shame the self-promoter. It
is preceded by much verbal taunting and badgering
from the other members of the hunting party.[21] The
sharing of meat is an aspect of this egalitarianism.
Meat is not necessarily doled out with an eye toward enhancing one's status. Instead, those who
are shared with seem to be in the driver's seat to
some extent, since they are able verbally to ha-
rangue the sharer and, in doing so, manipulate his
behavior. There is no alpha male in hunter-gatherer
society, no top-ranking man who dictates what the
group will do and where it will go. The sharing of
meat could lend itself to the emergence of an alpha
should the gift of protein and fat be considered by
the entire group to be socially coercive. Instead, the
acceptance of the gift can be somewhat manipulative. Perhaps what we see among traditional foraging people is something akin to what bonobos do.
Male bonobos defer to females when the female
wants food, but this is likely strategic deference in
exchange for mating opportunities. In the same
way, successful human hunters may defer to the

supplication of beggars not only to give away sur-
plus meat that is difficult to keep, anyway, but also
in order to exert the tacit influence suggested in
Kristen Hawkes's showing-off model.

In order to understand the place of hunting and
meat in the world of the forager, we should ask how
and why meat is shared. We saw that for the Aché,
sharing appears to be strategic on the part of those
who control the kill, benefiting them socially and
even reproductively if not necessarily nutritionally.
But when is sharing not just sharing?

Traditionally, anthropologists have viewed sharing by foragers as open-hearted, beneficent giving
intended to help others even at the expense of de-
priving oneself. Some anthropologists, however,
beg to differ. Bruce Winterhalder sees the origins of
human uniqueness in intragroup sharing patterns.[22]
Whether by scavenging or by hunting, early hominids obtained meat and may have used it as a currency for barter, as chimpanzees sometimes do.
When Aché hunters give meat to the families of
non-wives and end up having sexual affairs with
these women, is this not a trade of nutrients for sexual opportunities? When a male chimpanzee who
has captured a prized colobus meal offers it to a
sexually swollen female in exchange for a mating
with that female, it is also a trade of valued goods
for reproductive access. Such sharing, far from altruistic, may reflect evolved aspects of social manip-

ulation. Sharing may also have evolutionary roots in the concept of tolerated theft. If what you have is more than ample, then the cost of losing some of it to a persistent beggar may require less energy than the expense of trying to keep it. So surrendering a bit of your prize may be tactical avoidance. More-over, the apparent generosity of foragers may be matched by their persistence in soliciting. Of course, people trade goods for services all the time. In Western industrial society, those in positions of power are often able to use their status to negotiate terms of trade that are favorable to them. Dealing from strength is a common feature of a cash econ-omy. If someone needs to borrow money from you, you agree to make the loan contingent on a promise that it be paid back in full, plus interest. This inter-est is a reflection of the economic power of the lender.

But in foraging societies, egalitarianism and an apparent lack of hierarchy are the rules that people live by. Why do great apes and nearly all modern people live in rigidly hierarchical societies, while foragers do not appear to possess the concept of hi-erarchy? Christopher Boehm has considered this question and what it may mean for the way in which cultural mechanisms of change can influence biological evolution. He reasons that the prevention of hierarchy formation is enforced from those who are in nonleadership positions, but that leadership does exist, albeit controlled from beneath. Boehm refers to the manipulative use of the leader by his followers as a reverse dominance hierarchy and considers it to characterize most foraging societies. Through their conscious rejection of the dominance status of anyone who tries to tout his own accomplishments, foragers succeed in bringing the would-be alpha quickly back to earth, statuswise. Consensus is all, and in this striving to achieve consensus Boehm sees the roots of modern culture, able to override the biological evolutionary process.[23] David Erdal and Andrew Whiten also see the roots of our behavior in the food sharing of foragers, but in a different light. They consider food sharing to be a learned tradition that is grounded in an evolved disposition to share. They approach the forager food-sharing paradox from a biological perspective contrasted with Boehm's more cultural-determinist view. Erdal and Whiten suggest that at some point in human evolution, the cognitive ability to enforce a reverse dominance hierarchy arose, which in turn placed a great natural selection value on the ability to be politically astute in dealing with group mates.[24] How this astute political intelligence emerged and how it has directed the evolution of our behavior are the subject of the next chapter.

The Ghost in the Gorilla

How clever does a man or monkey
need to be before the returns on
superior intellect become vanishingly
small?

N. Humphrey,
"The Social Function of
Intellect" (1976)

Most features of our bodies, from our eyes to our
pumping hearts, have obvious purposes that can be
deduced by comparison with other animals. The
plain truth is, however, that we do not know why
we and our recent ancestors evolved such large,
complicated brains. The function may seem clear:
our brain is big and we are an abundant, successful
species. Therefore, the big brain must have granted
us our gift of survival. This may be true; the brain is
a most expensive organ to grow and nourish, and it
is hard to imagine its dramatic expansion occurring
unless it served some essential role in survival and
reproduction. That role remains a mystery. In this
chapter, I examine the evolutionary roots of human

cognition and suggest that they may be related to our hunting and, more important, meat-sharing ancestry. It is unlikely that brain expansion in the primate order was made possible or necessary solely by predation, since only a few of the big-brained primates are hunters. It may, however, be necessary to be big-brained to both hunt and make use of the spoils of the hunt in a strategic way.

An evolutionary approach to human cognition is a fairly recent phenomenon. The philosophical issues surrounding Descartes' split between mind and brain in humans has been described as the "Ghost in the Machine"; we know where the brain sits, but where and what is the mind? The history of the study of human cognition has been like the peeling of an onion, each layer of which is one more canon of divine intervention that falls beneath the weight of natural explanation. The outer layers came off early; the earth is not the center of the universe and humans are not preordained centerpieces of the evolution of life. At the onion's heart is the idea that the human psyche is simply an organic outgrowth of the evolution of the brain. Most of us have no trouble with the first claims, but many still have qualms about the center of that onion. Careers in the history of science have been built on the denial that a mind such as ours could occupy a brain so similar to that of an ape without some divine or at least quasi-supernatural influences. As the philosopher Daniel Dennett has suggested, many modern-day thinkers unconsciously invoke a supernatural explanation—a skyhook—for the origins of our intellect by failing to appreciate the Darwinian evolutionary aspects of the human mind. This chapter is premised on a Darwinian theory of mind that considers the psyche as organic an adaptation as any other aspect of our biological makeup, and applies that point of view to the cognitive aspects of hunting and meat sharing in humans and in nonhuman primates.

A bonobo sits down in front of a box containing treats that can only be gotten by cutting a thick length of twine that holds the box's door shut. After a moment's reflection, the ape picks up a nearby stone and hurls it at the ground (showing no interest in using the expected flaking method practiced by ancient humans). The rock breaks, exposing a razor edge. The bonobo then takes its new tool and methodically cuts through the cord, opening the door and accomplishing the required task. Kanzi, a bonobo who has been the subject of ape language studies conducted by Sue Savage-Rumbaugh, did his stone tool manufacture in an experiment conducted by archaeologist Nicholas Toth. Kanzi showed that he possesses enough foresight to put a stone to use in a novel way in order to obtain resources from his environment.[2] We know that early humans began making and using stone tools in much the same way about 2.5 million years ago. Charles Darwin had no way of knowing when tool manufacture originated, but in 1871 he wrote that tools had provided the first impetus to the evolution and elaboration of the human brain and cognition. In Darwin's view, early humans were freed from the use of their hands when they became upright walkers, later becoming proficient tool users.[3] Natural selection favored those individuals who were bright enough to make better and better tools for survival skills, and the end result was a big-brained, bipedal, manually dextrous hominid. Unfortunately for Darwin, we now know that bipedalism arose with the first hominids at least five million years ago. There is no solid evidence of stone tool use before 2.5 million years ago, so the transition from apes to hominids was not stone-tool driven. And the expansion of the brain's volume occurred even later, much less than one million years ago. Our greatest biological thinker had his insights in order, but he had a poor working knowledge of evolutionary time, which suggested to him a feedback loop that never existed.

A MIND IN THE FOREST

Each November, Gombe chimpanzees feast on a seasonal bounty of termites. The termite *Macrotermes* builds enormous earthen mounds that may stand two meters tall and extend an equal distance below the ground. As the dry season gives way to the first rains, the mounds become wet and muddy. At this time, tunnels are opened to the surface and kept plugged with mud when not in use by the soldier caste. Chimpanzees know this and spend hours each day investigating the many mounds in their territory, probing with their fingers into potential holes. On finding an open tunnel, the apes slide their termite fishing tools—twigs, stems, or grass blades fashioned into fishing poles—into the opening, where they are grabbed by the mandibles of the protective soldiers.[4]

During the season of termite foraging, chimpanzees sometimes travel from mound to mound all day long. A mound may be as near as 50 meters or as far as a kilometer from the next one, but the chimpanzees seem to know exactly where each termitary is located. I once followed Goblin, the formerly powerful alpha male, as he made his way between a dozen mounds in lower Kakombe valley. Midway between destinations, he stopped at a bush and broke off several small twigs. Using his incisor teeth he clipped them to the desired length, then ran one hand down their lengths to strip off their leaves. He then took his new tools and placed them in his mouth sideways, where they were carried in a beeline to the next mound. This is a clear example not only of having spatial memory to navigate between termite mounds but also of the foresight involved. Goblin did not wait to reach his tool-use site before selecting his tools; he was thinking about what he was going to be doing several minutes hence as he made his way there. He was sharpening his pencils for the exam awaiting him in the next room.

During long days in the field with chimpanzees, Gombe researchers and Tanzanian field assistants have long discussions about chimpanzee intelligence. The question is often: "How much do the chimpanzee know about their world?" We follow a chimpanzee party as it travels from fruit tree to fruit tree, seeming to know the precise location and the fruiting season of thousands of food sources. Is this entire data bank stored in the animals' memory, or are the locations of only some major food trees memorized, while they randomly bump into all other foods en route? To those who doubt that a chimpanzee could remember the location and the fruiting seasons of thousands of individual trees, consider your own shopping skills. You don't stroll down the aisles of your neighborhood supermarket hoping to catch a chance glimpse of the broccoli; instead you know where each item is, how much it costs, and in which months seasonal foods are likely to be in stock. Now imagine that the market is expanded in size to cover an entire square kilometer. Remembering the locations of thousands of items becomes much more difficult. But instead cf spending only an hour per week in the market, suppose you live there. From dawn to dusk every day of the year you search for oranges, pasta, and oatmeal. Neither chimpanzees nor we have great trouble remembering this sort of complex environment. Nor is this some sort of optimal foraging strategy that relies on an efficient travel pattern rather than spatial memory. Apes do not find food by foraging in a rigidly preset pattern any more than we mail a letter by walking in concentric circles until hitting a mailbox.

"Ecological intelligence" is the hypothesis that, as early hominids began to use increasingly complex environments, the brain's neocortex, or cerebrum, increased in volume and also underwent a reorganization to navigate the food resource labyrinth. In a lightly wooded African grassland mosaic, patchily distributed fruit trees might be farther apart than in the deep forest. Emerging hominids' large brains may have allowed them to navigate their way to distant food sources. There is good evidence for a connection between large neocortex size and the diet and pattern of food distribution among primates. Katharine Milton of the University of California, Berkeley, advocates the view that foraging necessities gave rise to primate brain size expansion. She showed that among the New World monkeys she studies, there is a direct correlation between eating high-quality, patchily distributed foods and brain neocortex size.[5] This correlation in turn may be evolutionarily related to the form and function of the species' social system. Barro Colorado Island is a small island sitting in the body of water created during the building of the Panama Canal. Both howler and spider monkeys traverse the island in search of food. Howlers eat a diet heavy in leaves, especially young tender leaves, which are relatively predictable in occurrence and uniform in distribution. Howlers are also among the most slothful of primates, slowly making their way to leafy banquets, where an entire group will feed in a closely knit assemblage. By day's end, the howler group will have spent little time and energy traveling and many hours eating and soporifically digesting.

The same forest is home to a population of spider monkeys, which are about the same size and proportions as howlers. Spider monkeys are, however, high-energy primates. They fission into small subgroups to travel long distances in search of the ripe fruit they prefer. Rather than eating the monotonous diet of leaves that howlers eat, spider monkeys prefer a diverse diet that includes mostly carbohydrate-packed ripe fruit of many species. Keeping track of the locations and the periods of availability of this supply is daunting. The other aspect of their biology that distinguishes spider monkeys from howlers is their relative brain-to-body weight ratios. Spider monkeys have one of the largest brains of the higher primates relative to their body weight. Howlers, meanwhile, are at the lower end of the anthropoid primates in the same measure; they have, instead, extra-large digestive tracts for a monkey of their size. The implication is that the spider's brain is the product of millions of years of natural selection, rewarding the ability of big-brained primates to remember the details of their physical environment. Howlers rely on a more sedentary lifestyle combined with careful food choice from a poor-quality resource base to cope with the exigencies of a very different diet.

You may have noticed something about the ecology of spider monkeys that I've just described. They are very chimpanzee-like in their diet, their foraging behavior, and their fission-fusion grouping patterns. And like chimpanzees, spider monkeys have very large brains relative to the size of their bodies. Milton suggests that when hominids evolved in a complex forest-grassland mosaic, a premium was placed on locating and remembering food sources. This resulted in larger and more powerful brains. Some researchers have pointed out that there are many forest animals that make a successful living amid the complexities of the rain forest or woodland-grassland without the benefit of a large brain. But having a big brain may be one response to natural selection pressure for foraging efficiency in complex environments.

The physical environment is only one of the milieus in which primates live. Go to the zoo and watch a group of monkeys or apes. Even in the most sterile and esthetically barren enclosures, the monkeys tend be psychologically healthy provided they have access to one feature: other monkeys. In the most depressingly bare cages, a cage mate is a saving grace. Of all the environmental influences that mold the behavior of a higher primate, none is greater than the social environment. A social creature faces a different set of selected traits than a nonsocial one, and a social being that lives in cohesive, long-lasting associations must have special capabilities. A group-living animal must be able to recognize other group members, cope with the constant competition that comes with group living, and coordinate its movements with the rest of the group. None of these requires a very impressive brain; fish live in schools and birds flock with unimpressive cognitive abilities. But living in social groups and remembering all the debts and credits of life is very different. This requires the ability to recognize the other members of one's group as individuals, and to keep a running scorecard of the other group members' relationships to you and to one another over a period of months or years. We know that nonhuman primates routinely perform such cognitive feats. Dorothy Cheney and Robert Seyfarth of the University of Pennsylvania showed that East African vervet monkeys are astute anthropologists who use and abuse social knowledge in voluntary, manipulative ways. Vervets cannot associate a python's obvious track on the ground with the possible presence of the python in the thicket to which the track leads. But the same vervets score high on social I.Q.: they recognize kin relationships within their group and utter alarm "words" that indicate which predator species is attacking them.[6] Psychologist Frans de Waal's landmark study of a zoo colony of chimpanzees showed that it helps to be clever and manipulative to rise in the dominance hierarchy, to curry favors from matriarchal females, and to navigate around the social icebergs that dot the landscape of a primate group's political web.[7]

Many animal behavior researchers attribute the increase in the size of our brain to increasing natural selection pressures that favored socially and politically adept group living. Primates that could best manipulate their social surroundings to their own advantage reap more mates and leave more offspring. This is of course also a time-honored pattern used by political leaders throughout human history. This school of thought—"Machiavellian intelligence"—has risen to prominence in the past decade, stemming from seminal writings in the 1960s and 1970s by scholars such as psychologist Nicholas Humphrey and primatologist Alison Jolly.[8] There can be no more basic goal of social life for an intelligent creature than to get one's fellow group members to do what you want them to do without knowing that they are being manipulated. Males enlist the help of other males to overthrow an alpha whom neither could hope to challenge alone. Once the alpha has been toppled, the fragile coalition of low-rankers may fall apart, each left to his own devices to hold onto the power that they both gained but could not successfully share.

Intentionally deceiving a colleague is another piece of evidence that animals are socially very smart. Lying is not a uniquely human trait. Cheney and Seyfarth saw vervet monkeys give false alarm calls, apparently to divert group mates' attention from food resources the group was trying to share.[9] When some primates copulate, the male (and sometimes the female) gives copulation calls—loud shrieks that attract attention from all around. Low-ranking male rhesus macaques do not give these calls as often as high-ranking ones, presumably to avoid being bashed by a high-ranking male who does not approve of the mating.[10] But low-ranking male chimpanzees have been seen trying manually to prevent their lips from parting in order to suppress copulation calls—an apparently conscious and forethinking attempt to deceive. I once saw a Gombe chimpanzee employ tactical deception as a means of gaining access to a female in the presence of the alpha. Beethoven, a physically impressive but low-ranking adult male, attempted to mate with a popular female, Gremlin. As a party of chimpanzees sat in a clearing, he employed one of the courtship gestures that male chimpanzees use to signal their expectations to a female—rapidly shaking a small bush in view of the female. But when Gremlin approached Beethoven, alpha male Wilkie interceded and drove Beethoven off. Beethoven stayed on the periphery of a cluster of chimpanzees in the clearing, then abruptly did an uncharacteristically bold charging display past Wilkie and Gremlin and into the undergrowth beyond. Alpha males do not take such insubordinate shows of machismo lightly. Wilkie followed immediately with his own much longer charging display, which carried him 20 meters away into a thicket, whereupon Beethoven strolled back to Gremlin and they copulated safely out of sight of the receding Wilkie.

THE brain increased incrementally in size throughout nearly all of primate evolution. Then, since the time of *Homo erectus* and the gradual transition into modern humans some 200,000 years ago, human brain size exploded. This explosion happened at a time when humans were probably living in small bands, leading a nomadic or seminomadic life of hunting and gathering. They were also probably evolving a qualitatively different and more advanced form of language and speech than had existed in earlier hominids. It may be that the brain size increase that took place during this time was due to the value of a larger and more sophisticated brain for survival and reproduction. As the complexity of human societies grew, so did the pressure that natural selection brought to bear on the ability to be socially and politically clever. The result is a species in which frequent small deceptions, planned maneuvers of one's mates, and general politicking skills count for more in nearly all arenas of life than do physical size, strength, or agility.

The roots of intelligence may lie in a combination of ecological complexities, the value of foresight in making and using tools, and the value of being socially intelligent. Teasing apart these factors is, however, a difficult practice when one is without fossilized soft tissue such as brains. British primatologist Robin Dunbar compared data from the life histories of a range of primate species to examine how the ecological function of intelligence stacks up against its political value.[11] He found that the size of the neocortex and the size of the social group in which the primate lives were highly correlated across the primate order. A monkey that lives in a group of fifty instead of a group of five must be able to hold in its head the knowledge of a much more complex web of social relationships. It is like

playing chess versus tic-tac-toe. Dunbar found no
relationship between brain size and the size of the
home range used by a primate, suggesting that the
physical environment's complexities had not led to
big brains. Nor did the extent to which the species
used "extractive foraging" (tool use) to obtain food
correlate with cortex size. Dunbar's findings suggest that while the first push toward a larger brain
may have been the result of a patchily distributed,
high-quality diet and the cognitive mapping capabilities that accompanied it, the evolution of the
very large brain of the higher primates was primarily due to its value in social intellect (fig. 6.1). There
are many confounding factors here, since lower primates such as prosimians usually live in small
groups while monkeys and apes reside in bigger
groups. This may simply reflect their evolved predispositions for particular grouping patterns.

There is another factor that is ignored by most
advocates of the social intelligence school of thought.
Although brain sizes can be measured, plotted, and
correlated against the mating system and grouping
pattern, whether higher primates really have much
more complicated social lives is less clear. Does the
raw number of potential social relationships dictate
a more complex social life? We assume so but have
few pieces of empirical support for the idea. This is
important in considering how the modern human
brain came to reach its current size. The modern

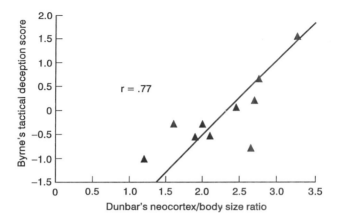

FIG. 6.1. The relationship between Dunbar's brain-to-body-size ratio and Byrne's index of tactical deception. The species in which tactical deception is most often seen also have the relatively largest neocortex of the brain. (Adapted from Byrne 1995)

human brain is on average three and a half times larger in volume than a chimpanzee's (1400 cm³ compared to 400 cm³). But what evidence is there that early human societies were more complex than modern chimpanzee societies? If group size is the primary factor, then how do group sizes compare between foraging people and chimpanzees or bonobos? The evidence is weak, since we cannot assume that ancient humans lived in bands that were comparable to those of modern human foragers. There is wide variation in group size among modern foragers and also among chimpanzees. Chimpanzee communities can have over one hundred members; this is larger than most bands of modern foraging people, although chimpanzee communities are not cohesive and members are rarely all in the same place. The !Kung, for example, spend most of their lives in small camps of only fifteen to twenty people. If the increase in primate brain size through evolution has social intellect as its root, then we must be able to explain the dramatic increase in brain size in the very recent past of *Homo sapiens* based on a measurable increase in social complexity.

> I knew at a glance that this was no ordinary anthropoidal brain. Here was the replica of a brain three times as large as that of a baboon, and considerably bigger than that of any adult chimpanzee.[12]

Perhaps in the final analysis, brain size does not matter most. The relationship between brain size and intelligence has long been a conundrum over which researchers have argued. Harry Jerison first established the evolutionary relationship between the size of the brain and that of the body in which it resides, calculating an index of this ratio as a measure of evolutionary intelligence.[13] But Richard Byrne has argued insightfully that we must distinguish between the value of a sheer increase in the number of neurons through evolution—the biological equivalent of building bigger and bigger computer disk drives—and the influence of body size. Larger bodies may require more neurons per unit of brain volume to handle the neural traffic required to run them; the analogy is of an old telephone switchboard. Alternatively, the brain could function like the Internet, with the critical functions far less localized than either a disk drive or a switchboard. Neural messages are continually rerouted along myriad switching points, and it is not clear how important sheer size increase would be compared to more efficient organization.[14] Terry Deacon argues that we are badly mistaken when we try to delineate a smoothly progressive increase in brain size through primate evolution, because brain-size-to-body-size ratios measured in otherwise very different animal groups are not comparable. Deacon shows that much of the variation in brain-to-body size among primates may be due to natural selection operating on body size rather than natural selection producing a certain, larger brain size.[15] He accuses the bigger-is-always-better advocates of perpetuating an obsolete evolutionary ladder concept of brain size, one that is both methodologically and logically flawed. We would not say that just because an animal has bigger eyes or longer legs, all animal species could be ranked based on visual acuity or running speed. In the same way, we should not be trying to do so based on the relative size of a species' brain. Whatever the key influences have been on primate brain size, the much more recent expansion of the human neocortex is the most difficult to account for. The selection factor may have been language, which is without fossil evidence and is therefore in the realm of informed speculation.

Puzzling out the evolution of human cognition has long been fair game for evolutionary thinkers. In the 1970s, anthropologists Robin Fox and Lionel Tiger described, in a popular book, *The Imperial Animal*, the psychological underpinnings for human group social life using interpretations based on our nonhuman primate past.[16] In a second book, Tiger also expounded on related ideas of male cooperation and the evolutionary advantages of it.[17] Tiger and Fox argued that modern humans employ behavioral strategies whose predispositions were molded in earlier eras in our evolutionary history. If we want to understand why we choose a particular mate or we strive to climb the corporate ladder of success, we must pay attention to our evolutionary past. More recently, a new school of thought has reworked these ideas by borrowing from cognitive psychology and information science, and called itself evolutionary psychology. Evolutionary psychologists have set out to explain the Darwinian basis of human behavior and located its roots in a theoretical complex of psychological mechanisms they call cognitive domains. These domains are modules through which behavior is mediated, and they are argued to be the ultimate objects of natural selection. When an adolescent boy finds himself attracted to every woman around him except his own sisters, that is because the incest avoidance domain is at work. The reason we behave irrationally in some cases—killing ourselves by fighting to achieve a position of status—is the desirability in our ancestors of achieving notoriety, perhaps because in an earlier time it was accompanied by acquiring more mates. According to John Tooby and Leda Cosmides of the University of California, Santa Barbara, the roots of such behaviors are to be found in the psychological seeds of humanity that evolved in the Pleistocene, and which, in the short time that we have lived in modern societies, natural selection has had insufficient time to modify.[18] So we are, according to evolutionary psychologists, fundamentally out of touch with the environment in which our social behavior evolved, which has led to many modern social and psychological problems.

In their search for evidence that modern people operate on a cognitive plane shaped by a long history of natural selection, evolutionary psychologists have erred in their level of analysis. There is no reason to consider the cognitive domains by which we respond to our social environment to be uniquely human. Nearly all of what evolutionary psychologists call the adapted human mind is actually the adapted ape mind carried into the present by the continued benefits of avoiding incest, of mating well, and a host of other behaviors. The arbitrary choice of the Pleistocene for the environment to which the mind is adapted ignores the fact that the roots of human cognition must be sought much deeper in our past, when our nonhuman primate ancestors had to respond to many of the same social contexts. I suspect that evolutionary psychologists are reluctant to admit that there is an evolutionary psychology of gorillas as well as of humans. Treating the environment in which our psychological mechanisms evolved as "cave man days" allows any maladaptive behavior of modern people to be dismissed as simply adapted to a different time and place. A second problem with evolutionary psychology lies in the lack of rigor and the overaburdance of philosophy that many evolutionary biologists perceive in the paradigm. The cognitive domains we presume to exist may lie beyond hypothesis testing. While the psychological aspects of the evolution of our own cognition should be explainable in Darwinian terms, the models developed to deal with this last bastion of mystery are still a vague, quasi-philosophical endeavor.

Intelligence and Hunting

August 4, 1953. Dusk is falling over Lake Tangaryika and I am returning from Busambo point, where I spent the afternoon looking through a telescope into Lower Mkenke valley, censusing a group of red colobus monkeys. At about 6 P.M., I watch the gathering sunset and then head down the trail toward the beach. As I reach the bottom of the trail I run into (literally) a party of male chimpanzees walking up the trail past me—Frodo, Freud, Beethoven, Tubi, and Gimble, plus others whose movements I hear in the nearby thickets. As they pass me they turn onto a side trail that parallels Mkenke stream, heading toward the area where my colobus should be settling down for the night. I turn and follow, and minutes later we meet the colobus group in low trees on the valley slope. The chimpanzees charge up the trees while the colobus alarm-call and flee frantically. Instead of grouping for a joint defense against their attackers, most of the colobus cluster in the crown of a tall oil palm tree, where they sit peering over the fronds at the chimpanzees ascending the trunk below. With male colobus above and the chimpanzees blocked from view below, the hunters give up and withdraw, and the hunt appears to be over only a few minutes after it has begun.

But Frodo has a different plan in mind. He climbs the palm that holds the colobus until he is just below the crown. He cannot see the colobus well, and they probably cannot see him at all. After trying to peer up and over the crown unsuccessfully, Frodo gently pulls one frond down, so that its length bridges the gap between the palm and the adjacent tree. The colobus see the newly formed bridge to safety, but not the hairy hand beneath it. One by one they begin to run and leap across the frond and into the next tree. One after another they cross over, and I (concealed behind a bush some distance away) can see Frodo looking up at each colobus that passes safely. After seven or eight have crossed, a mother carrying her young infant on her abdomen attempts the crossing. As she runs across the frond, Frodo releases the frond and lunges at the female and baby. His grab attempt is slightly too short and too late, and mother and baby flee to safety. Frodo has pursued a clever strategy for catching a mother or an infant or both, and this time narrowly missed.[19]

Whether intelligence is an important factor in the hunting tactics of chimpanzees is a key issue because of the putative importance of hunting to the evolution of human intelligence. If intelligent hunters succeed more often than less intelligent hunters, and if the capture of meat has some survival value to the animals, then natural selection should have favored those chimpanzees who employ clever hunting tactics. If females prefer to mate with the best hunters, perhaps because the females benefit nutritionally from gifts of meat, then hunting performance would be subject to sexual selection as well. Since little evidence exists that Gombe chimpanzees set off each morning with the intention of finding meat, optimizing foraging routes to take advantage of the likely locations of colobus is probably not part of the hunting strategy. Many animal species, from bumblebees to hummingbirds, forage efficiently without any semblance of higher intelligence, since natural selection may program the ability to optimize travel routes to hit a maximum number of quality food sources per unit of area.[20] Certainly predatory behavior involves little higher cognition for many animals. A leopard needs extraordinarily evolved sensory capabilities and the weaponry to stalk and kill prey, but not the higher cognitive function that lies at the basis of human brain evolution. In many mammalian predators, however, hunting tactics and the expected behavior of the prey species must be learned. This is true for solitary hunters like leopards as well as social hunters like lions and wolves. Scavenging behavior, for instance, might not be subject to strong natural selection pressures in chimpanzees given that scavenging is both rare and purely opportunistic. But even scavenging would have provided natural selection with fertile ground to enhance the spatial memory and navigational abilities of early humans. Skills that might be important to making a kill, such as the flexibility to respond strategically to rapidly changing circumstances, should also be selected for and enhanced.

Among chimpanzees and a few other primates, eating other animals is a regular part of life. Although the traditional view has been that few pri-

mates hunt, predation is in fact widespread in the
primate order if we include the capture of smaller
animals such as amphibians, rodents, and other
small vertebrates. Among the higher primates,
those species that regularly eat animal protein are
the same species for which other higher cognitive
attributes have been reported. Among New World
monkeys, for example, some populations of capuchin monkeys are voracious predators of small animals. They are omnivores that, depending on the
habitat, invest up to 70 percent of their waking
hours engaged in animal protein foraging. Capuchins are also among the most skilled of extractive
foragers. They spend hours rummaging intensively
through leaf litter on the ground and bromeliads
high in the canopy for whatever small animals they
can find. In the course of doing so, they continually
catch small animals as their hands toss aside dead
leaves, palm fronds, and bits of the forest floor.
Many of these prey are flushed using sticks to probe
and poke into unreachable places. Adult capuchins
weigh only about 4 or 5 kilograms (8–10 pounds,
but this does not prevent them from hunting animals that weigh a sizable percentage of their own
weight, such as squirrels. Brown capuchins in
southeastern Brazil may even prey upon the squir-
rel-sized and tamarin monkeys that live in the same
forests![21]

Lisa Rose and Susan Perry have both studied the

predatory behavior of white-faced capuchins in a
seasonally dry forest in Costa Rica, and their observations of hunting behavior make for interesting
comparisons with chimpanzees.[22] The monkeys kill
prey about twice a week, approximately equal to
the rate of kills by chimpanzees at Gombe. As in
Gombe chimpanzees, males do most of the hunting,
and hunting is more frequent in the dry season than
during rainy months. Again like chimpanzees,
high-ranking males are not necessarily the most
skilled hunters, but they are the most able to appropriate carcasses from other hunters and so control
more meat than other group members. Most kills
are either squirrels or the infants of raccoonlike
coatis. Coati babies are taken from their nests in
large numbers during the coati breeding season.[23] In
the hunting of coatis and squirrels, however, an essential difference is revealed between capuchin and
chimpanzee hunting patterns. Capuchins hunt
coatis by stealthily approaching the nests of these
mammals, snatching the babies from the nest and
making off with them. When hunting squirrels, capuchins give chase and pursue the rodents singly, in
pairs, or occasionally in what appears to be a tag-
team relay, with one hunter dropping out as another picks up the pursuit. There is little evidence of
cooperation during the hunt. In the end, the carcass
of the squirrel is rarely shared by the captor among

the hunters who participated, or with any other group member.[24]

In addition to their extractive foraging and meat-eating facilities, capuchin monkeys share another trait with chimpanzees: their very large brain-to-body size ratio. To what extent might this correspondence between capuchin and chimpanzee hunting be linked to their high degree of neocortex expansion? Although both are higher primates, chimpanzees and capuchins are distantly related and their brain size similarity is only partly due to a common ancestry. Both species must cope with a complex environment in which food resources are patchily distributed. But many other smaller-brained primates that do not eat meat have the same problems. Primatologist Linda Fedigan has suggested that capuchins are good hunters because they have extractive foraging skills that are useful in catching animal prey. They do this in a complex habitat containing inedible animals that the capuchins must avoid eating by mistake. She points out that male capuchins may hunt more often because they are larger relative to their squirrel prey and because females are often burdened by dependent offspring, making hunting difficult and perhaps dangerous.[25] These same factors apply to the preponderance of male hunting among chimpanzees. Lisa Rose has extended this reasoning to compare capuchins with the high-level extractive foraging skills employed by chimpanzees when procuring termites and biting ants from their nest through the use of tools.[26]

There are, however, other correlates of large brains and meat eating that suggest that substantial cognitive attributes are necessary to be a strategically clever meat consumer and sharer. Humans, chimpanzees, and capuchins are at the top of the scale both in cerebral volume measures and also in being politically smart. This suggests either that some higher cognitive skills are involved in hunting or that it helps to have such skills in making use of the meat afterward. We have already seen that chimpanzee cultures of hunting and strategies of sharing vary from one forest to another. Gombe chimpanzees hunt in groups with selfish nepotistic goals at stake, while at Taï hunting is more cooperative. Christophe Boesch has claimed that highly cooperative hunting is a sign of advanced intelligence, because it requires coordinating the actions of two or more individuals in pursuit of a common goal. However, many predators cooperate when hunting without the benefit of a highly elaborated neocortex. It may be unfair to consider cooperation a more advanced state of intelligence than selfish manipulation, deception, and nepotism, as much as we humans would like to think otherwise. We need to consider the behavior of meat eaters not while they are hunting but rather when they are dividing the spoils to see evidence of cognition that distinguishes big-brained primates from other animals.

COGNITION, COOPERATION, AND THE ART OF THE DEAL

The recent history of primate behavior study has seen a tidal change in the way we look at animal cognition. Scientists who attributed to chimpanzees the sorts of individual differences that we routinely ascribe to humans were once derided as anthropomorphizers. Those days are happily gone and the current generation of researchers takes into account the distinctive attributes of each member of a primate group. As the social intelligence school of thought rose to the fore, researchers began to describe how individuals use their power and privilege to control others, to obtain more resources for themselves, and to be astute observers of their own neighbors. Among the behaviors that are of most interest is food sharing. This is due to the cooperation and reciprocal altruism that sharing entails. In the face of much evidence that animals tend to act in their own self-interest first and foremost, we try to explain why sharing exists and how it could evolve. From a Darwinian perspective, the sharer ought to reap more than he invests, for without that selfish impetus he would not share. The difficulty in imagining the evolution of sharing among both humans and other animals is that such cooperative behavior is difficult to establish. Sharing is especially difficult to maintain in the face of those who selfishly cheat. If one individual cheats the system by reaping the benefits of sharing without contributing, that cheater will benefit. If sharing is at all genetically influenced, the cheaters will end up leaving more offspring in the next generation, and the cheating trait would spread, leading to the collapse of cooperative sharing. An example from my own work on antipredator defense makes this clear. Six or seven male red colobus monkeys mount a cooperative defense when attacked by chimpanzees bent on eating them. This is a risky business, since males are occasionally killed for their heroism. In the midst of a chimpanzee-colobus battle, I have seen male colobus watching the action from a safe perch a bit distant to the fray. By failing to codefend, the bystander reduces his own risk of injury while allowing his group mates to bear that risk. He is cheating the system. Why some males would stay back from the counterattack while others leap into it is unclear; perhaps those males who have reason to think they have fathered babies in the group have a reproductive incentive to protect those progeny. Males with no such incentive might instead seek to protect themselves in hopes of mating more in the future.[27]

In the pattern of sharing we see the roots of both sides of our altruistic and selfish natures. The art of the deal among sharers is to give a bit less than you receive. Giving just enough to perpetuate the relationship with the fellow sharer is often the goal. This subtle form of cheating is common among sharers of many primate species, including humans. Whether it is subject to penalty or correction is largely determined by the relative status of the giver and sharer. For example, I borrow fifty dollars from you today and promise to pay you back next week. But next week arrives and I say I do not have the money, though I am able to pay back twenty-five dollars. You must then decide whether to take the 50 percent refund and risk that you may never see the rest of your money, or hold out and insist on the full amount. Your economic need will determine how much I must pay back. If you need the money badly enough right now, you will be forced to take half payment; I have put you on the short end of the transaction due to your lower financial status. This is the situation in which a subordinate member of the chimpanzee dominance hierarchy may find himself. He is unable to negotiate from a position of strength but needy of whatever social alliances he can get, no matter how one-sided. In such instances sharing may appear reciprocal to the casual observer when in fact the powerful are manipulating others to their advantage. For example, in the recent history of the Kasakela chimpanzee community at Gombe, the alphas have not necessarily been highly skilled or avid hunters. They are always, however, avid meat eaters. The dominance status of the alpha male allows him to appropriate carcasses with impunity from other hunters. The others, though more skilled at capture, are less able to defend their catch due to their socially subordinate position. At both Gombe and Mahale, sharing is nepotistic and the sharer reaps political rewards, while at Taï sharing of meat is more often reciprocal among the hunt participants. These two sharing traditions, one Machiavellian and the other more reciprocal, have apparently arisen in response to aspects of the local relationship between chimpanzee and colobus. Christophe Boesch has argued that at Taï the enormous height of the rain forest canopy to which the colobus retreat makes cooperation in hunting essential, with the result that helping out is rewarded later. At Gombe, with a lower forest structure and greater ease of hunting, such cooperation is not necessary and so no such reward system has developed.[28]

Once a capuchin hunt has ended, opportunities arise to share and be shared with. Nearly all food-sharing observed in Lisa Rose's study was between adults and infants; adults very rarely shared with adults. This is in contrast to what we see among chimpanzees and bonobos. Although most food-sharing among apes is between mothers and their infants, sharing among adult males and females is also very much a part of many posthunt sharing fests. In captive studies, capuchins will share generously with cage mates, even passing food through the bars to another monkey without the possibility of having the favor returned.[29] But in the wild, capuchins are not liberal in their sharing relationships with others, even when the captor possesses a chunk of meat too large for him to eat quickly himself. Likewise, Shirley Strum's baboons, discussed in chapter 2, are not generally liberal sharers; most possessors of meat simply try to monopolize it for themselves.[30]

Humans, chimpanzees, baboons, capuchins: is there a common thread linking these hunting, meat-loving primates? All four species rank among the highest in the frequency of tactical deception and other forms of social, political intelligence. Humans, chimpanzees, and capuchins are three of the largest-brained primates relative to the size of their bodies. All four species have relatively large prey species available: red colobus for chimpanzees, gazelle fawns for Gilgil baboons, squirrels for capuchins, and a wide variety of small and large prey for early hominids. Chimpanzees and capuchins are among the most skilled tool users of all nonhuman primates; indeed, the two show the most advanced extractive foraging skills of any animal apart from ourselves. It seems unlikely that this co-incidence of traits that accompanies hunting and meat eating is entirely random. I suggest that the ability to make use of meat for nutritional purposes is facilitated in a social primate by a relatively high degree of intelligence, because of the complexities of sharing the meat of other animals. Being a sharer requires a level of encephalization that is seen only among the great apes and humans.

SHARING AND THE ROOTS OF HUMAN COGNITION

Sharing and egalitarian relationships are two hallmarks of the foraging people of the modern world. The open reciprocity that occurs routinely in these human societies is rare among nonhuman primates. There is a stark contrast between the egalitarian nature of hunter-gatherer bands and the strongly hierarchical structure of the great-ape societies. Yet when we consider modern people in virtually all industrialized societies, we see status and hierarchy as the rule of life. David Erdal and Andrew Whiten of the University of St. Andrews in Scotland have depicted the evolution of hierarchy as a U-shaped curve, with the early rise of hierarchical society followed by a drop in status seeking among early humans.[31] Hierarchical behavior returns only recently among modern people, perhaps when subsistence foragers became settled as agriculturalists and began living in larger groups.

In the egalitarian relationships that exist among foragers, we may see the shaping of the origins of the human mind. Alliance formation, always important among nonhuman primates, assumes even greater weight among humans because of the value in cooperating to obtain meat. Such coalitions in chimpanzees allow males jointly to challenge those much higher in the dominance hierarchy. Christopher Boehm of the University of Southern California suggests that in the course of human evolution, the political savvy that stems from the need to form alliances contributed to the rise of human intelligence. In humans, the ability to destabilize dominance hierarchies was the basis for the system of dominance reversal that characterizes hunter-gatherers.[32] Egalitarianism is a system in which the self-interests of the individual are at times submerged in order to benefit the group. There is no leader; the impulse to lead or to achieve social status is beaten down by verbal harassment from everyone else. In the same way, the behavior of foraging people strongly discourages getting too much of anything at the expense of the other, including too large a share of meat after a kill is made. So the goal is not to get ahead oneself, but to ensure that no one else gets ahead. If the social intelligence required to navigate a social hierarchy is considerable, the political cleverness required to live within a nonhierarchical system may be even greater. One must be acutely aware at all times of one's own needs, gains and losses, and be able to compare them to those of many other comrades. Erdal and Whiten see this process leading to increasing plasticity of behavior, both enabled by and leading to a larger and more powerful brain. In the behavioral options available to a social strategist in an egalitarian setting, we may see the foundations for modern human cognition. The paradox is that within the past several thousand years we have seen a return to hierarchical societies, complete with chieftains, social castes, and social institutions based on social or economic status. Nearly all these systems are male-dominated patriarchies. In patriarchal systems, males control essential resources and use them to control females and reproduction. In the final chapter, I will consider the roots of meat eating and meat sharing and their link to the evolution of patriarchies.

Meat's Patriarchy

Women's social standing is roughly
equal to men's only when society
itself is not formalized around roles
for distributing meat.

R. Leakey and R. Lewin,
People of the Lake (1978)

Hunting was not about getting
enough vitamin B 12.

D. Haraway,
Primate Visions (1989)

*A male chimpanzee in a Tanzanian forest is chewing on
the limb of a monkey carcass that he has caught. Other
members of the hunting party cluster around him at-
tempting to get a morsel. Juveniles sit on the forest floor
below, hoping that fragments of bone or drops of blood
will fall to them. The beggars in the tree ask for meat
with outstretched, upturned palms, and place their fin-
gertips at or even inside the hunter's lips in order to get
his attention. He tries to ignore their presence, continually turning his body away from them. One exception to*

*this indifference is a female, who carries a sexual swell-
ing and receives the male's attentions. He allows her to
take pieces of the carcass, which she then shares with her
infant. Occasionally she presents her swelling to him
and they copulate. Meanwhile the possessor of the meat
allows one other chimpanzee, a longtime ally, to take bits
of meat also. This interplay of meat with sex, political
networks, and status displays is typical of the strategic
meat-sharing pattern seen among chimpanzees.*

Man the Hunter argued in part that human brain
size and cognitive skills were enhanced through a
long evolutionary history pursuing prey, with all
the communication and coordination skills that
hunting may place at a premium. This is partly correct; social hunting does place a premium on intelligence. The premium is exacted, however, in the
realm of the sharing, control, and distribution of
meat after a kill rather than in the pursuit of the
prey. In the doling out of the meat we see signs of
strategizing and politicking that go far beyond
those seen in predatory behavior. The evolutionary
legacy of our hunting and scavenging past lies
therefore not so much in the hunt but in the division
of spoils. While there appears to be little continuity
in the hunting tendencies of ancient and technologically simple modern people, there is a vital link in
the use of meat as a currency, a valued good. This
link is found in the behavior of apes and people

when the sharing of meat occurs, or when food is used to manipulate the behavior of others. In the distribution of meat in apes, as in other social arenas, we see control and power at the heart of both male and female patterns of behavior. In the attempts by either sex to tip the balance of power, we may be seeing the roots of human gender relations. When meat becomes a resource that is not only a food but also a social currency—a way to help you obtain what you want in the group—we are seeing the emergence of barter, currency-based human social systems.

In human and some other primate societies, meat eating is about politics as well as nutrition. The control of a valued resource is about power. When the two sexes are involved in the power struggle, the physically dominant sex often controls a resource and therefore controls female reproduction as well. Sexual politics plays a key role in chimpanzee meat eating just as it does in some traditional human societies. Relationships between the sexes are part and parcel of the capture and sharing of meat, since in nearly all human and nonhuman primate societies in which the meat of mammals is eaten frequently, males hunt for it more often than females. We see gendered carnivory in human, ape, and monkey societies. In chimpanzee society, both males and meat occupy a hierarchy: males dominate one another, and all males dominate all females. Meat is so highly desired that the whole community will devote hours to catching it, even though most female and immature community members end up with only tiny scraps. No other food commands such devotion. The hierarchy of meat appears to be closely linked to the hierarchy of males, in that it is almost always males that capture meat, putting them in the role of providers for other community members, including females.

In many traditional human societies, men hunt but women procure most of the protein and calories for their social groups through their gathering of roots, fruits, and small animals. This behavior has led some anthropologists to claim that the importance of hunting and meat eating was more mythical than real, since men's efforts, while it received much attention from a previous generation of researchers, did not account for much nutritionally. But this misses the key point about meat eating. The fact that meat is so highly valued even when it composes a small part of the diet is powerful testimony to its value as social currency. Men are able to use meat to enhance status, show beneficence, and even to obtain more sex by having caught meat. Nutrition is not irrelevant to meat's value at all; it is essential in that it makes the meat something worth bragging about, begging for, and manipulating with. But the use of meat has gone well beyond this; the !Kung do not sit and haggle for hours over the ownership of a pile of berries, but they will do so over the carcass of an antelope. The value of meat is a matter of perception by group members. It is irrelevant that plant foods are as *valuable* a resource as meat. What matters is that animal carcasses are considered by both men and women as a more *valued* resource than plant foods.

This difference in the way that plant foods and meat are regarded in both ape and human societies may be related to the difficulty of obtaining meat, and that only one sex—typically males—controls meat. A fundamental premise of the origins of patriarchy is that when men gain control over resources that women need, they use it to control and coerce them. Male dominance is a near cultural universal; this is accepted by both feminists and sociobiologists, as well as by scholars who are both. This does not mean that all societies are in all ways patriarchal; women have spheres of influence that may simply differ from the arenas in which men rule. There are some societies, including those of the traditional foragers, which are more egalitarian than others. There are, however, no true matriarchies. Feminist scholars usually depict male domination of women as arising in historic times, due to the social contexts of our own recent prehistory. Few, however, acknowledge the possibility that male dominance may have far more ancient origins. Perhaps this is because if patriarchy has its roots in the

Miocene or Pliocene instead of the past few thousand years, it implies an evolved Darwinian basis of patriarchal systems. Among other primates, male dominance is usual, though not universal. Where it occurs, it is based on a combination of greater male size and strength, male coalitions, and a lack of strong female-female alliances.

In all primate species, including ourselves, there are sexual conflicts of interest over male desires for access to reproduction. Females become not the goal of patriarchal domination itself, but rather the means to its end, which is reproductive success. This evolutionary view of human gender relations is quite different from a feminist perspective, which more often considers the goal of male domination to be an end in itself.[1] It was once thought that women were subject to male control in most cultures because of limitations imposed by their reproductive biology. Women were thought to be intrinsically barred from ruling due to their physiology. This antiquated view still finds an audience with those who question whether women can serve as combat soldiers or governmental leaders due to their menstrual cycles. Modern feminist thinkers have seen the control of reproduction as the male goal in human societies.[2] Female reproductive biology may limit women from playing controlling roles in some societies, but this is due more to male oppression than to women's lack of ability or desire to rule. This analysis has not often been properly grounded in the study of nonhuman primate societies. If we ask how patriarchies came into being in the first place, a feminist approach might be to suggest males' greater physical size and strength. Some psychological need to dominate might also be invoked. But any biologist knows that male size and strength are not an end, but a means to an end, namely the control of female bodies and their reproductive output. Competition among males, and female choice of male mates with particular physical features, led to the 10–15 percent size difference between ancestral human males and females.

Two biological anthropologists who are both feminists and sociobiologists, Sarah Hrdy and Barbara Smuts, have been exceptional in offering biological levels of explanation for human gender relations. In separate studies, both suggest that the same Darwinian principles that govern the conflicting interests of the sexes in all primate social systems may prescribe male coercive behavior toward females in humans. They point out the logical inconsistency of feminist analyses that posit the same rationales for male dominance in an attempt to explain how such systems came into being, without asking why they exist at all. The "why" of patriarchal systems can be addressed with the extension of feminist theory to evolutionary paradigms. This may seem an unlikely wedding. The social and historical explanations for male domination offered by many feminist thinkers are simply a recent manifestation of the deeper biological process that has created patriarchal social systems in our close relatives. Some recent authors have recognized that sociobiological analyses of gender relations, far from offering sexist, simplistic views, reveal the complexities in human behavior. Females are not passive receptacles of male reproductive desires. They are active strategists in pursuit of their own interests and often are the driving forces behind the social system itself. Males may likewise be caregivers rather than warriors and headmen. However, each sex has reproductive interests at stake in social life, and these may clash. There is no direct genetic basis for patriarchy; they arise from the biological conflict of interests between males and females of all primates, including men and women.[3]

If patriarchies are about male dominance and the control of female reproduction, then we should consider how males acquire power. This may point to the role that meat eating and meat sharing plays in patriarchal systems. One way this occurs in nonhuman primate groups, human social groups, and probably in our ancestors' social groups, is through sheer size and strength. Sexual dimorphism is the product of sexual selection favoring male attributes; in species where being large helps in the mating game, we see outsized males with impressive canines and musculature. The level of dimorphism in modern human populations is about 10 percent in stature and somewhat more in weight. This difference is very similar to the magnitude of size difference between males and females in both chimpanzees and bonobos, but vastly less than that seen between the sexes among gorillas and orangutans. The degree of size dimorphism suggests the intensity of male competition for mates, but the fit is not always clearly tied to the social system. The largely solitary orangutan, for instance, shows the greatest degree of male-female size dimorphism, exceeding 50 percent. Males may sexually coerce and brutalize females.[4] Females are not passive, and may refuse to mate with a male no matter how imposing he is, though she may subject herself to violent attack by doing so.

A second way that males obtain power is through strategic alliances. This is true for both human and nonhuman primate societies. Male coalitions may form to control access to females, or to prevent other males from having access to them, or to defend territories that contain desired females. Anthropologists Joseph Manson and Richard Wrangham have theorized that primate societies featuring tight male bonds form when the crucial resources are both portable and defensible—such as females—rather than immovable, such as fruit trees.[5] Chimpanzees, bonobos, and many tribal human

societies fit this dichotomy. When it is highly beneficial for males to form alliances, there is often a kin-
selected benefit that heightens the potential rewards for cooperating with kin.

The converse effect of male alliance formation is
that in many female primate species, females do not
form strong alliances. This is especially true among
the great apes, in which female alliances are weak
in three of the four great apes. Because female apes
usually disperse from their natal groups at sexual
maturity, they end up living as breeding adults in a
new group lacking relatives and close allies. Some
degree of bonding among females may develop
over time, but females that disperse to new groups
usually do not have a support system akin to that of
males remaining in their natal groups. Barbara
Smuts has pointed out that this lack of alliance formation contributes to the formation and maintenance of patriarchies by preventing females from
bonding for purposes of protecting themselves
from male ambitions.[6] Females may, however,
choose not to form alliances with other females because to do so would be against their own reproductive interests. Smuts suggests that a female's reluctance to form coalitions to the degree that males
do may be based on their own selfish reproductive
interests.

The history of gender relations has not been entirely Nature red in tooth and claw. Men and

women have long worked toward common goals as reproductive partners, group members, and as parents. However, both biological and social/historical perspectives lead us to believe that each sex is also carrying out a more selfish individual agenda. And in both the evolutionary and social interpretations of patriarchal systems, meat eating has played an important role.

The way in which meat is distributed among the members of a social group, and the way in which gender roles may be related to meat-sharing patterns, was a preoccupation of our discussion of chimpanzee behavior. The same gender-biased acquisition and distribution pattern that characterizes these apes also characterizes many human groups. Meat is a difficult-to-obtain resource in both human and many other primate societies, valued out of proportion to its nutritional worth. And as a resource much in demand by both sexes but typically brought home by males only, it becomes an important factor in the efforts of males to influence the behavior of females, and vice versa.

In her book *The Sexual Politics of Meat*, feminist writer Carol Adams considers the role of meat eating in human societies in shaping perceptions of women. She observes that women in many societies eat a diet that is mainly plant foods while men in the same societies eat more meat. The delegation of a second-class diet to women is one aspect of a patriarchal society.[7] Peggy Sanday surveyed a range of traditional societies and concluded that those that eat a diet in which meat plays an important part are more likely to be strongly patriarchal.[8] In some traditional societies, the rules of meat division preclude women from getting as much as men, particularly nutrient-rich portions such as the fat in a carcass.[9] Meat thus has a gendering influence. Adams, in analyzing the distribution of meat and power in human societies, makes essentially the same argument that any ape or human sociobiologist would make regarding meat and patriarchy, but couches her argument entirely in historical and social terms. Adams points out that when males control a resource as highly valued as meat, its worth as a nutrient is largely mythical. Tubers and beans make an equally protein-rich diet. But men eat meat in many cultures in the belief that it gives them the strength that they need for their work. Women are not deemed to require it. Meat is also a nearly universal symbol of masculinity, from Western industrial to forager societies, and the eating of meat is thought to enhance masculinity. The image of a vegetarian weightlifter or football linebacker is paradoxical to most of us—evidence that these values are deeply entrenched in Western culture. Perhaps this is because meat eating is associated with meat getting and the other masculine attributes traditionally connected to hunting.[10]

This chain of logic allows Adams to characterize the value given to meat in mythical terms. She considers meat to be a sinister symbol of a male-dominated world, rooted in patriarchal myths rather than reality. The idea that patriarchy's foundations are to be found only in mythically ascribed traits combined with male dominance does not, however, fit well with biological realities. Instead of a sinister symbol of the cultural influences on patriarchy, meat eating is more likely a behavioral focus around which patriarchies have evolved. Meat has long been a symbol of masculinity only because it served males well throughout human evolution as a political currency that is used to enhance male alliances, snub rivals, and control females. Ironically, Adams and other feminist scholars who write about patriarchy may be factually correct about the recent history of male behavior toward women. By choosing to ignore the evolutionary level of analysis in the prehistory of male behavior toward females, their analyses are ultimately shortsighted. Only a biological interpretation of the relationship of meat to male dominance can fully address the roots of patriarchal societies. This is not to say that meat eating is the central feature of all male-dominated societies. It can, however, be a defining trait of gender relations in some societies, such as those in which men hunt for a living while women gather.

This brings us back to Man the Hunter. The idea

that meat eating may have been a catalyst in human behavioral evolution fell into disfavor due to the backlash against Man the Hunter. The model was pilloried to the extent that it became a forbidden term, one that labeled its adherents as male-biased chauvinists, consciously or unconsciously ignorant of the role of women in human evolution.[11] This backlash was fueled by key field data about the role of women and also fit emerging social values that sought to place women in an equable role in society relative to men. Whether the actual role of females in early human prehistory fits our modern Western gender-balanced sensibilities is an entirely different question. While women may collect most hunter-gatherer protein, we should not ignore the fact that men are able to use meat for their own selfish and manipulative political ends.

When a male chimpanzee withholds a scrap of meat from a female until she mates with him, we see the use of meat as a manipulative social tool. The value of the resource to the female may be entirely nutritional, but its value to the male is clearly both social and nutritional, in that he may be able to use it as a means to a political and reproductive end. In the same way, the value of a carcass to a lactating woman may be entirely different than its worth to a man. Both ape and human scenarios involve the control of a resource, valued perhaps beyond its amino acids or calories.

If females usually live in male-dominated, male resource-controlled groups, how do they obtain what they want in those groups? Female primates, including humans, may need alliances and strategic support to survive and reproduce. One female goal in any species is the successful birthing, rearing, and nurturing of offspring. The successful maturation of offspring is a goal in which males are equally interested but in which they tend to invest far less energy and time. Males thwart female ambitions, and they often do so in ways that involve the control of reproduction. Female primates can navigate the complex social web of relationships, male power struggles, and their own power struggles only by being politically clever. The premium placed on social intelligence in females extends to obtaining food, allies, and mates. Gagneux and colleagues' recent study of furtive mating outside the community by female Taï chimpanzees is a vivid example of strategic reproductive behavior.[12] Obtaining valued food is also strongly influenced by a female's ability to network within her group. Being fertile, high-ranking, and clever are three important ways that females can obtain meat. Since no female is fertile for more than a small portion of her life, and only a few females can be high ranking, most must rely on their ability to manipulate their social milieu, including the males themselves, at the same time that males are trying hard to manipulate them.

Life in a primate society is thus about power and control, not of females by males, but of each sex by the other. Each side has a valued resource to gain and also has much to lose.

The relationship between hunting and sex is nowhere clearer than among the Sharanahua, a people of the Peruvian Amazon basin. According to anthropologist Janet Siskind, "Put at its crudest, the special hunt symbolizes an economic structure in which meat is exchanged for sex." (p. 103). Since women produce as much food as men do, this is not likely to be just an economic exchange in which women exchange what they produce for what men produce. It is a culturally mediated system, with an obvious biological effect, in which sex is the incentive for men to hunt, and men who are better hunters have a better chance to have wives or mistresses. Siskind considers women to be a scarce commodity to be competed over in the Sharanahua, because sex is not free for men and must be won, and because some men have more than one wife, enhancing the impetus to "win" women with proffers of meat. This form of barter may be a common occurrence among tropical forest people.[13]

This control game is played out in many nonhuman primate societies as well as in our own human cultures. Machiavellian intelligence applies to both males and females, but it does not necessarily apply to them in the same ways. Males and females have often been likened to separate species, with separate reproductive strategies and different means of meeting the same end of reproducing themselves.[14] They may also have evolved somewhat different sorts of cognition as a result of needing to cope differently with complex social environments. Certainly males and females live in separate social worlds even in the same social group in many species. The same claim has been made about men and women.

How does the social intelligence of females play itself out in the real world? Men control access to scraps of a carcass. Meat is a resource that females want, though perhaps not for the same reasons it is desired by males. There is no reason to think that for several individuals sitting around a fire, meat holds the same promise. For a woman it may mean essential amino acids to nourish her fetus or infant, but to a man it may signify an ability to demonstrate his beneficence or his hunting ability by sharing it with her. He may seek an alliance with her that is either politically strategic or reproductively beneficial, or both. She, meanwhile, has her own set of self-interests about control, reproduction, and political power. She may not, however, have unconstrained access to a resource as valued as meat. So she must wheel and deal to get it, just as the male is wheeling and dealing to secure her as a mate, partner, or link in his political support chain. Far from the passive secondary role in which some early models of human behavioral evolution portrayed her, these females actively pursue their own ends, and also achieve their own spheres of power and influence, that affect both male control strategies and the social dynamics of the entire group.

We receive conflicting messages in our society about the value of sharing. While a deep cultural value is given to altruistic sharing, we are also socialized to be individualistic. We learn that there are both powerful incentives and disincentives to share and to be selfish. Either can be advantageous depending on the context; altruists can be strategically self-serving in myriad ways, and individuals who are otherwise selfish may seem utterly altruistic at times. If these are qualities that served us in our distant ancestry, then perhaps that is why there is so much ambivalence about altruism and selfishness in society today.

I have portrayed the roots of human behavior as manipulation and social cunning that arise from the use of meat in our ancestors. This is very different than saying that, because of a meat-eating past, we have an innately aggressive nature. It should be remembered that predators, while possessing many adaptations that can be used in aggression within their own species, do not necessarily use them. In fact, many scholars doubt whether the predatory aggression that allows a lion to bring down a zebra or a chimpanzee to savagely rip apart a monkey is even closely related to intraspecific aggression. Aggressive behavior within a species may resemble predation, but the resemblance is often superficial. Humans are not demons by nature; in spite of the attention that we focus on human violence, there are thousands of acts of compassion for every act of physical aggression.

The hunting, scavenging, and sharing of meat were fundamental features of the lives of our ancestors. This does not mean that we are biologically driven to do any of these. The way that we deal with one another in society is rooted in social strategies that were molded during a time in our history when getting and using meat was vital. If meat were a currency with a 10,000-generation history in the human family, then the traditions that have developed related to the use of meat are likely to have some evolved basis. By sharing meat we are both altruistic and selfish, as we are in most other arenas of our endeavors. We are not simply compassionate by training and Machiavellian by nature. Nor are we constrained by our past to repeat Machiavellian patterns in the future. A fuller understanding of our ancestral nature is, however, the first step to liberation.

Notes

CHAPTER 1
THE INDELIBLE STAMP

1. Stanford and Allen 1991; we were responding to models/ scenarios of human evolution developed in, among others, Tooby and DeVore 1987; Foley and Lee 1989.

CHAPTER 2
MAN THE HUNTER AND OTHER STORIES

1. Kingston and colleagues (1994) show that Pliocene hominid habitats in East Africa were likely a woodland and woodland/ grassland mosaic rather than savanna with only scattered trees.

2. Moore 1996.

3. Stanford and Allen 1991.

4. Wrangham and Peterson 1996.

5. Zihlmann et al. 1978.

6. Latimer et al. 1981.

7. For example, Tooby and DeVore 1987; Foley and Lee 1989. Both decry the use of a chimpanzee analogic model, but in the end they conclude that our last common ancestor was very chimpanzee-like.

8. Hill 1994.

9. Wrangham 1987.

10. Ghiglieri 1987.

11. McGrew 1979, 1981.

12. Isaac 1978.

13. Strum 1981, 1983.

14. Connor et al. 1992.

15. Rodman and McHenry 1980.

16. The volume resulting from the conference was *Man the Hunter* (1968), edited by Richard Lee and Irven DeVore.

17. Washburn and Lancaster 1968. See Cartmill (1993) for a discussion of Man the Hunter as a fable.

18. Simmons and Scheepers (1996) suggest that the tradtional Darwinian explanation for neck length in giraffes—longer necks are for feeding—is not supported well by field data.

19. Tanner and Zihlmann 1976.

20. Murdock 1965.

21. Fedigan 1986.

22. Wrangham 1980, for example.

23. See Gero and Conkey 1991.

24. Morgan 1982.

25. Langdon (1997) shows that most of the supposed adaptations to aquatic life that Morgan invokes in human anatomy and physiology are actually shared with a wide variety of nonaquatic mammals.

26. Steudel 1994, 1996.

27. Taylor and Rowntree 1973.

28. Rodman and McHenry 1980.

29. Jablonski and Chaplin 1993.

30. Dart 1953.

31. Falk 1990; Wheeler 1984.

32. Hunt 1994.

33. Rose 1984.

34. Lovejoy 1981.

35. Washburn 1959.

36. Steudel 1994, 1996.

37. However, a recent discovery (Köhler and Moyá-Solá 1997) of a 9 million-year-old ape fossil that appears to have been bipedal has shed new light on the origin of human posture. Living on an island in the Mediterranean region, *Oreopithecus bambolii* had a modified form of bipedal posture and locomotion that may have evolved in parallel with the hominid bipedality that was to arise several million years later in Africa. The authors speculate that the island habitat, which had no predator species, may have

afforded a safe, isolated environment in which upright posture could have evolved as a foraging adaptation.

38. Washburn merged his troglodytian theory of human origins with knuckle-walking models of human origins, and both Washburn and the ideas held sway for many years (Tuttle 1974).

39. But see Boesch and Boesch 1989, who argue that Taï chimpanzees at times (mainly in the hunting season) search for colobus monkeys.

40. Aiello and Wheeler 1995.

CHAPTER 3
APE NATURE

1. Caccone and Powell 1989.

2. Hill and Ward 1988. Although there is some debate, Miocene fossils from China and Pakistan assigned to the genera *Sivapithecus* and *Ramapithecus*, are frequently considered to be ancestral orangutans.

3. Nishida 1968. In Mahale National Park, 100 kilometers south of Gombe on the eastern shore of Lake Tanganyika, Nishida and a team of primarily Japanese scientists studied chimpanzees for nearly as long as Goodall and her students have worked at Gombe (see Goodall 1968). The Japanese were the first to outline the social system of the chimpanzee, referring to what Western scientists call the community as the "unit-group." There is a fascinating convergence between Western and Japanese approaches to the study of primate societies, whose history is described in D. Haraway's *Primate Visions* (1989).

4. Gagneux et al. 1997.

5. Stanford et al. 1994a. Gombe chimpanzees kill up to seven colobus at a time, though this is rare, and one or two kills is the norm. The number of kills is directly correlated with the number of hunters. In a hunt at Gombe in 1992, seven colobus were killed, totaling nearly 30 kilograms of carcass weight. See also Teleki 1973.

6. Boesch and Boesch 1989; Boesch 1994.

7. Nishida et al. 1992.

8. Goodall 1986; Stanford et al. 1994a,b.

9. Stanford 1998a; Stanford 1996.

10. Kano 1992; White 1988.

11. Furuichi 1987; see also Idani 1991.

12. Stanford 1998b.

13. Tutin 1979; see also Wallis 1997.

14. Kano 1996.

15. Takahata et al. 1996. Mahale male chimpanzees copulated at a higher mean rate than male bonobos at Wamba (*P. troglodytes* 0.20–0.29/hr; *P. paniscus* 0.10–0.21/hr). These rates are for adults, but adolescent male chimpanzees also have higher copulation rates than adolescent male bonobos. Bonobo mating rates are thus not higher than those among chimpanzees if the swelling duration and not the entire interbirth period is used as the time frame. If the entire interbirth period is included, then bonobos show higher copulation rates, because the period of maximal swelling occupies a slightly larger percentage of the menstrual cycle in bonobos than in chimpanzees. In their earliest field accounts of bonobo sex, researchers were cautious about interpreting differences with chimpanzees (see Thompson-Handler et al. 1984).

16. De Waal 1987, and implications thereof in Wrangham 1993.

17. Kano 1992.

18. Wood and White 1996.

19. Fossey 1983; Schalle 1963.

20. Kuroda et al. 1996; see also Goldsmith 1996; Remis 1997; Tutin 1996; Yamagiwa et al. 1996; Jones and Sabater Pi 1971.

21. Watts 1989.

22. Mitani 1985; see also van Schaik and van Hooff 1996 for a review of orangutan behavior and ecology.

23. Wrangham 1980; van Schaik and van Hooff 1983.

24. Moore 1984a; Strier 1994.

25. Stewart and Harcourt 1987.

26. Manson and Wrangham 1991.

27. These captured monkeys sometimes ended up dead from

seemingly unintentional mistreatment during handling. Sabater Pi et al. 1993.

28. Sugardjito and Nuhuda 1981.
29. Stanford 1998a.
30. Hohmann and Fruth 1993; also Ihobe 1992.
31. Milton and Demment 1989.
32. Kano 1992.

CHAPTER 4
THE VIEW FROM THE PLIOCENE

1. Sept 1992, 1994.
2. Shipman and Walker 1989.
3. Yellen 1991.
4. Dart 1953.
5. Ardrey 1961.
6. Brain 1981.
7. Binford 1981.
8. Speth 1991.
9. Stiner 1991.
10. Potts and Shipman 1981; Bunn and Kroll 1986; Shipman 1986.
11. Blumenschine 1987; but see Tappen 1995 for a counter-argument.
12. Blumenschine 1986.
13. Marean 1989.
14. Schaller and Lowther 1969.
15. Isaac 1978; Isaac and Crader 1981.
16. Bunn and Ezzo 1993.
17. Boesch 1991.
18. Speth 1987, 1989.
19. Speth and Davis 1976.
20. McGrew 1992.
21. Stanford 1995.
22. Plummer and Stanford, submitted manuscript.
23. Caro 1994.
24. Cheney and Seyfarth 1991.

25. Plummer and Stanford, unpublished manuscript.
26. Howell 1966.
27. Binford 1987; see also Binford 1981.
28. Binford 1987.
29. Stiner 1991.
30. Trinkhaus 1987.
31. Straus 1987.

CHAPTER 5
THE HUNTING PEOPLE

1. Lee 1979. A debate has long raged about how influenced the !Kung San have been by the outside world.
2. Bailey et al. 1989.
3. Smith and Winterhalder 1992, for example.
4. The description of this Aché day is taken from Hill and Hawkes 1983.
5. Rodman and McHenry 1980; see related discussion in chapter 3.
6. Although note that hunter-gatherers such as the Aché often kill small prey without using weapons. Even after shooting a monkey out of a tree with a bow and arrow, the kill is sometimes made by hand (Hill and Hawkes 1983).
7. Lee 1979.
8. O'Connell and Hawkes 1988; Bunn and Ezzo 1993.
9. Alvard 1993.
10. Speth 1991.
11. Stanford 1995
12. Hill and Hawkes 1983; Kaplan and Hill 1985a.
13. Hill and Hawkes 1983.
14. Hawkes 1991.
15. Kaplan and Hill 1985a.
16. Ibid.
17. Blurton-Jones 1984.
18. Hawkes 1991; see also Moore 1984b; Hawkes 1993.
19. Kaplan and Hill 1985b.
20. A /Xai/xai San hunter, in Lee 1979, 244.

21. Lee 1979.

22. Winterhalder 1996, 1997.

23. Boehm 1993, and in press.

24. Erdal and Whiten 1994, 1996.

CHAPTER 6
THE GHOST IN THE GORILLA

1. Dennett 1995. A skyhook is a *deus ex machina* invoked consciously or unconsciously by those who accept the fact of evolution but are skeptical about the role that natural selection has played in the origins of the mind.

2. Toth et al. 1993.

3. Darwin 1871. In *The Descent of Man*, Darwin has the facts right but the chronological sequence, and therefore the feedback loop, wrong.

4. See McGrew 1992 for an extensive discussion of chimpanzee tool use and its implication for early hominid evolution.

5. Milton 1981.

6. Cheney and Seyfarth 1981.

7. de Waal 1982.

8. Jolly 1966; Humphrey 1976.

9. Cheney and Seyfarth 1991.

10. Hauser 1993; Manson 1996.

11. Dunbar 1992.

12. Dart 1959, 5.

13. Jerison 1973.

14. Byrne 1995.

15. Deacon 1990.

16. Tiger and Fox 1971.

17. Tiger 1969.

18. Tooby and Cosmides 1992.

19. This anecdote obviously includes some interpretation of events: I *infer* that Frodo could see the colobus but they could not see him, and I *infer* that Frodo was using considerable foresight in letting the colobus see a bridge to safety that was actually a very clever trap set to catch one of them. I also *infer* that there

was a real connection between Frodo's actions and the actions taken by the colobus; that is, he was employing a strategy rather than just holding the palm frond in such a way that it appeared to be a trap. This is the sort of anecdote that may not fit into any empirical framework for interpretation. Anecdotes can only be retold, discussed, and perhaps videotaped; the underlying motivations and cognitive abilities that are suggested can be accepted or rejected.

20. Stephens and Krebs 1986.
21. Dietz, pers. comm. 1996.
22. Perry and Rose 1994.
23. Ibid.; Rose 1997.
24. Fedigan 1990.
25. Ibid.
26. Rose 1997.
27. Stanford 1998a.
28. Boesch 1994.
29. de Waal et al. 1993.
30. Strum 1981, 1983. The Gilgil baboons hunted more often than any other nonhuman primate population that has been recorded, but this may have been due to the influence of particularly avid males.
31. Erdal and Whiten 1994, 1996.
32. Boehm 1993; see also Boehm 1997.

Chapter 7
Meat's Patriarchy

1. Ortner and Whitehead 1981, and Lerner 1986, for example.
2. Martin 1992.
3. Smuts 1995; Hrdy 1997.
4. Smuts 1995; Smuts and Smuts 1993.
5. Manson and Wrangham 1991.
6. Smuts 1995.
7. Adams 1990.
8. Sanday 1981.
9. Speth 1990.

10. Adams 1990.
11. Tanner and Zihlmann 1976, for example.
12. Gagneux et al. 1997.
13. Siskind 1973.
14. Hrdy 1981, for example.

References

Adams, C. J. 1990. *The Sexual Politics of Meat*. New York: Continuum Publishing Company.

Aiello, L. C., and P. Wheeler. 1995. The expensive-tissue hypothesis. *Current Anthropology* 36: 199–221.

Alvard, M. 1993. Testing the "Ecologically Noble Savage" hypothesis: Interspecific prey choice by Piro hunters of Amazonian Peru. *Human Ecology* 21: 355–387.

Ardrey, R. 1961. *African Genesis: A Personal Investigation into the Animal Origins and Nature of Man*. New York: Dell.

Bailey, R. C., and N. Peacock. 1988. Efe pygmies of Northeast Zaïre: Subsistence strategies in the Ituri Forest. In *Coping with Uncertainty in Food Supply*, ed. I. de Garine and G. Harrison, 88–117. Oxford: Oxford University Press.

Bailey, R. C., G. Head, M. Jenike, B. Owen, R. Rechtman. and E. Zechenter. 1989. Hunting and gathering in tropical rain forests: Is it possible? *American Anthropologist* 91: 59–82.

Binford, L. R. 1981. *Bones: Ancient Men and Modern Myths*. New York: Academic Press.

Binford, L. R. 1987. Were there elephant hunters at Torralba? In *The Evolution of Human Hunting*, ed. M. H. Nitecki and D. V. Nitecki, 47–105. New York: Plenum Press.

Blumenschine, R. J. 1986. Carcass consumption sequences and the archaeological distinction of scavenging and hunting. *Journal of Human Evolution* 15: 639–659.

Blumenschine, R. J. 1987. Characteristics of an early hominid scavenging niche. *Current Anthropology* 28: 383–407.

Blurton-Jones, N. 1984. A selfish origin for human food sharing: Tolerated theft. *Ethology and Sociobiology* 4: 145–147.

Boehm, C. 1993. Egalitarian society and reverse dominance hierarchy. *Current Anthropology* 34: 227–254.

Boehm, C. 1997. Impact of the human egalitarian syndrome on Darwinian selection mechanics. *American Naturalist* 150: s100–s121.

Boehm, C. In press. *Hierarchy in the Forest.* Cambridge, Mass.: Harvard University Press.

Boesch, C. 1991. The effects of leopard predation on grouping patterns in forest chimpanzees. *Behaviour* 117: 220–241.

Boesch, C. 1994. Cooperative hunting in wild chimpanzees. *Animal Behaviour* 48: 653–667.

Boesch, C., and H. Boesch. 1989. Hunting behavior of wild chimpanzees in the Taï National Park. *American Journal of Physical Anthropology* 78: 547–573.

Brain, C. K. 1981. *The Hunters or the Hunted?* Chicago: University of Chicago Press.

Bunn, H. T., and E. M. Kroll. 1986. Systematic butchery by Plio/Pleistocene hominids at Olduvai Gorge, Tanzania. *Current Anthropology* 27: 431–452.

Bunn, H. T., and J. A. Ezzo 1993. Hunting and scavenging by Plio-Pleistocene hominids: Nutritional constraints, archaeological patterns, and behavioural implications. *Journal of Archaeological Science* 20: 365–398.

Byrne, R. W. 1995. *The Thinking Ape.* Oxford: Oxford University Press.

Caccone, A., and J. R. Powell. 1989. DNA divergence among hominoids. *Evolution* 43: 925–942.

Caro, T. 1994. *Cheetahs of the Serengeti Plains* Chicago: University of Chicago Press.

Cartmill, M. 1993. *A View to Death in the Morning.* Cambridge, Mass.: Harvard University Press.

Cheney, D. L., and R. M. Seyfarth. 1981. Selective forces affecting the predator alarm calls of vervet monkeys. *Behaviour* 76: 25–61.

Cheney, D. L., and R. M. Seyfarth. 1991. *How Monkeys See the World*. Chicago: University of Chicago Press.

Clark, W. Le Gros. 1967. *Man-apes or Ape-man?* New York: Holt, Rhinehart, and Winston.

Connor, R. C., R. A. Smolker, and A. F. Richards. 1992. Two levels of alliance formation among male bottlenose dolphins (*Tursiops sp.*). *Proceedings of the National Academy of Sciences* 89: 987–990.

Dart, R. 1953. The predatory transition from ape to man. *International Anthropological and Linguistic Review* 1: 201–219.

Dart, R. 1959. *Adventures with the Missing Link*. New York: Harper and Brothers.

Darwin, C. 1871. *The Descent of Man and Selection in Relation to Sex*. London: J. Murray.

Deacon, T. W. 1990. Fallacies of progression in theories of brain-size evolution. *International Journal of Primatology* 11: 193–236.

Dennett, D. C. 1995. *Darwin's Dangerous Idea*. New York: Simon and Schuster.

de Waal, F.B.M. 1982. *Chimpanzee Politics*. Baltimore: Johns Hopkins University Press.

de Waal, F.B.M. 1987. Tension regulation and nonreproductive functions of sex in captive bonobos (*Pan pariscus*). *National Geographic Research Reports* 3: 318–335.

de Waal, F.B.M., L. M. Luttrell, and M. E Canfield. 1993. Preliminary data on voluntary food sharing in brown capuchin monkeys. *American Journal of Primatology* 29: 73–78.

Dunbar, R.I.M. 1992. Neocortex size as a constraint on group size in primates. *Journal of Human Evolution* 20: 469–493.

Erdal, D., and A. Whiten. 1994. On human egalitarianism: An

evolutionary product of Machiavellian status escalation? *Current Anthropology* 35: 175–178.

Erdal, D., and A. Whiten. 1996. Egalitarianism and Machiavellian intelligence in human evolution. In *Modelling the Early Human Mind*, ed. P. A. Mellars and K. R. Gibson, 139–150. Cambridge, U.K.: McDonald Institute for Archaeological Research.

Falk, D. 1990. Brain evolution in Homo: The "radiator" theory. *Behavioral and Brain Sciences* 13: 333–381.

Fedigan, L. M. 1986. The changing role of women in models of human evolution. *Annual Reviews of Anthropology* 15: 25–66.

Fedigan, L. M. 1990. Vertebrate predation in *Cebus capucinus*: Meat-eating in a neotropical monkey. *Folia Primatologica* 54: 196–205.

Foley, R. A., and P. C. Lee. 1989. Finite social space, evolutionary pathways and reconstructing hominid behavior. *Science* 243: 901–906.

Fossey, D. 1983. *Gorillas in the Mist*. Boston: Houghton Mifflin.

Furuichi, T. 1987. Sexual swelling, receptivity, and grouping of wild pygmy chimpanzee females at Wamba, Zaïre. *Primates* 28: 309–318.

Gagneux, P., D. Woodruff, and C. Boesch. 1997. Furtive mating in female chimpanzees. *Nature* 387: 358–359.

Gero, J. M., and M. W. Conkey, eds. 1991. *Engendering Archaeology: Women and Prehistory*. Cambridge, U.K.: Blackwell.

Ghiglieri, M. P. 1987. Sociobiology of the great apes and the hominid ancestor. *Journal of Human Evolution* 16: 319–357.

Goldsmith, M. 1996. Ecological influences on the ranging and grouping behavior of western lowland gorillas at Bai Houkou in the Central African Republic. Ph.D. diss. State University of New York, Stony Brook.

Goodall, J. 1968. Behaviour of free-living chimpanzees of the Gombe Stream area. *Animal Behaviour Monographs* 1: 163–311.

Goodall, J. 1986. *The Chimpanzees of Gombe: Patterns of Behavior*. Cambridge, Mass.: Harvard University Press.

Haraway, D. 1989. *Primate Visions*. New York: Routledge.

Hauser, M. D. 1993. Rhesus macaque copulation calls: Honest signals for female choice? *Proceedings of the Royal Society of London (B)* 254: 93–96.

Hawkes, K. 1991. Showing off: Tests of an hypothesis about men's foraging goals. *Ethology and Sociobiology* 12: 29–54.

Hawkes, K. 1993. Why hunter-gatherers work: An ancient version of the problem of public goods. *Current Anthropology* 34: 341–361.

Hill, A. 1994. Early hominid behavioural ecology: A personal postscript. *Journal of Human Evolution* 27: 321–328.

Hill, K., and K. Hawkes. 1983. Neotropical hunting among the Aché of eastern Paraguay. In *Adaptive Responses of Native Amazonians*, ed. R. B. Hames and W. T. Vickers, 139–188. New York: Academic Press.

Hill, A., and S. Ward. 1988. Origin of the Hominidae: The record of African large hominoid evolution between 14 My and 4 My. *Yearbook of Physical Anthropology* 31: 49–83.

Hohmann, G., and B. Fruth. 1993. Field observations on meat sharing among bonobos (*Pan paniscus*). *Folia Primatologica* 60: 225–229.

Holmberg, A. R. 1950. *Nomads of the Long Bow: The Siriono of Eastern Bolivia*. Institute of Social Anthropology, Publication 10. Washington, D.C.: Institute of Social Anthropology.

Howell, F. C. 1966. Observations on the earlier phases of the European Lower Paleolithic. *American Anthropologist* 68: 88–201.

Hrdy, S. B. 1981. *The Woman That Never Evolved*. Cambridge, Mass.: Harvard University Press.

Hrdy, S. B. 1997. Raising Darwin's consciousness: Female sexuality and the prehominid origins of patriarchy. *Human Nature* 8: 1–49.

Humphrey, N. 1976. The social function of intellect. In *Growing Points in Ethology*, ed. P.P.G. Bateson and R. A. Hinde, 303–317. Cambridge, U.K.: Cambridge University Press.

Hunt, K. D. 1994. The evolution of human bipedality: Ecology and functional morphology. *Journal of Human Evolution* 26: 183–202.

Huxley, L. 1900. *Letters of Thomas H. Huxley*. Vol. 1. London: MacMillan.

Idani, G. 1991. Social relationships between immigrant and resident bonobo (*Pan paniscus*) females at Wamba. *Folia Primatologica* 57: 83–95.

Ihobe, H. 1992. Observations on the meat-eating behavior of wild bonobos (*Pan paniscus*) at Wamba, Republic of Zaïre. *Primates* 33: 247–250.

Isaac, G. L. 1978. The food-sharing behavior of proto-human hominids. *Scientific American* 238: 90–108.

Isaac, G. L., and D. C. Crader. 1981. To what extent were early hominids carnivorous? An archaeological perspective. In *Omnivorous Primates*, ed. R.S.O. Harding and G. Teleki, 37–103. New York: Columbia University Press.

Jablonski, N. G., and G. Chaplin 1993. Origin of habitual terrestrial bipedalism in the ancestor of the Hominidae. *Journal of Human Evolution* 24: 259–280.

Jerison, H. J. 1973. *Evolution of the Brain and Intelligence*. New York: Academic Press.

Jolly, A. 1966. Lemur social behavior and primate intelligence. *Science* 153: 501–506.

Jones, C., and J. Sabater-Pi. 1971. Comparative ecology of *Gorilla gorilla* (Savage and Wyman) and *Pan troglodytes*

(Blumembach) in Rio Muni, West Africa. *Bibliotheca Primatologica* 13: 1–95.

Jerison, H. J. 1973. *Evolution of the Brain and Intelligence*. New York: Academic Press.

Kano, T. 1992. *The Last Ape*. Stanford, Calif.: Stanford University Press.

Kano, T. 1996. Male rank order and copulation rate in a unit-group of bonobos at Wamba, Zaïre. In *Great Ape Societies*, ed. W. C. McGrew, L. F. Marchant, and T. Nishida, 135–145. Cambridge, U.K.: Cambridge University Press.

Kaplan, H., and K. Hill. 1985a. Food sharing among Aché foragers: Tests of explanatory hypotheses. *Current Anthropology* 26: 223–246.

Kaplan, H., and K. Hill. 1985b. Hunting ability and reproductive success among male Aché foragers: Preliminary results. *Current Anthropology* 26: 131–133.

Kelly, R. L. 1995. *The Foraging Spectrum*. Washington, D.C.: Smithsonian Institution Press.

Kingston, J. D., B. D. Marino, and A. Hill. 1994. Isotopic evidence for Neogene hominid paleoenvironments in the Kenya rift valley. *Science* 264: 955–959.

Köhler, M., and S. Moya-Solá. 1997. Ape-like or hominid-like? The positional behavior of *Oreopithecus bamboiii* reconsidered. *Proceedings of the National Academy of Sciences* 94: 11747–11750.

Kuroda, S., T. Nishihara, S. Suzuki, and R. Oko. 1996. Sympatric chimpanzees and gorillas in the Ndoki Forest, Congo. In *Great Ape Societies*, ed. W. C. McGrew, L. F. Marchant, and T. Nishida, 71–81. Cambridge, U.K.: Cambridge University Press.

Langdon, J. H. 1997. Umbrella hypotheses and parsimony in human evolution: A critique of the Aquatic Ape Hypothesis. *Journal of Human Evolution* 33: 479–494.

Latimer, B. M., T. D. White, W. H. Kimbel, and D. C. Johanson. 1981. The pygmy chimpanzee is not a living missing link in human evolution. *Journal of Human Evolution* 10: 475–488.

Leakey, R., and R. Lewin. 1978. *People of the Lake: Mankind and Its Beginnings*. New York: Doubleday.

Lee, R. 1979. *The !Kung San: Men, Women and Work in a Foraging Society*. Cambridge, U.K.: Cambridge University Press.

Lee, R. B., and I. Devore, eds. 1968. *Man the Hunter*. Chicago: Aldine.

Lerner, G. 1986. *The Creation of Patriarchy*. New York: Oxford University Press.

Lovejoy, O. W. 1981. The origin of man. *Science* 211: 341–350.

Manson, J. H. 1996. Rhesus macaque copulation calls: Reevaluating the "honest signal" hypothesis. *Primates* 37: 145–154.

Manson, J., and R. W. Wrangham. 1991. Intergroup aggression in chimpanzees and humans. *Current Anthropology* 32: 369–390.

Marean, C. 1989. Sabertooth cats and their relevance for early hominid diet. *Journal of Human Evolution* 18: 559–582.

Martin, E. 1992. *The Woman in the Body: A Cultural Analysis of Reproduction*. Boston: Beacon Press.

McGrew, W. C. 1979. Evolutionary implications of sex differences in chimpanzee predation and tool use. In *The Great Apes*, ed. by D. A. Hamburg and E. R. McCown, 441–464. Menlo Park, Calif.: Benjamin/Cummings.

McGrew, W. C. 1981. The female chimpanzee as a human evolutionary prototype. In *Woman the Gatherer*, ed. F. Dahlberg, 35–73. New Haven: Yale University Press.

McGrew, W. C. 1992. *Chimpanzee Material Culture*. Cambridge, U.K.: Cambridge University Press.

Meggitt, M. J. 1962. *Desert People: A Study of the Walbiri Aborigines of Central Australia*. Chicago: University of Chicago Press.

Milton, K. 1981. Distribution patterns of tropical food plants as a stimulus to primate mental development. *American Anthropologist* 83: 534–548.

Milton, K., and M. Demment. 1989. Features of meat digestion by captive chimpanzees (*Pan troglodytes*). *American Journal of Primatology* 18: 45–52.

Mitani, J. C. 1985. Sexual selection and male orangutan long calls. *Animal Behaviour* 33: 272–283.

Moore, J. 1984a. Female transfer in primates. *International Journal of Primatology* 5: 537–589.

Moore, J. 1984b. The evolution of reciprocal sharing. *Ethology and Sociobiology* 5: 5–14.

Moore, J. 1996. Savanna chimpanzees, referential models and the last common ancestor. In *Great Ape Societies*, ed. W. C. McGrew, L. F. Marchant, and T. Nishida, 275–292. Cambridge, U.K.: Cambridge University Press.

Morgan, E. 1982. *The Aquatic Ape*. New York: Stein and Day.

Morin, P. A., J. Wallis, J. Moore, R. Chakraborty, and D. S. Woodruff. 1993. Non-invasive sampling and DNA amplification for paternity exclusion, community structure, and phylogeography in wild chimpanzees. *Primates* 34: 347–356.

Murdock, G. P. 1965. *Culture and Society*. Pittsburgh, Pa.: University of Pittsburgh Press.

Nishida, T 1968. The social group of wild chimpanzees in the Mahali Mountains. *Primates* 9: 167–224.

Nishida, T., T. Hasegawa, H. Hayaki, Y. Takahata, and S. Uehara. 1992. Meat-sharing as a coalition strategy by an alpha male chimpanzee. In *Topics in Primatology*, vol. 1, ed. T. Nishida, W. C. McGrew, P. Marler, and M. Pickford, 159–174. Tokyo: University of Tokyo Press.

O'Connell, J. F., and K. Hawkes. 1988. Hadza hunting, butchering, and bone transport and their archaeological implications. *Journal of Anthropological Research* 44: 113–161.

Ortner, S. B., and H. Whitehead, eds. 1981. *Sexual Meanings: The Cultural Construction of Gender and Sexuality.* Cambridge, U.K.: Cambridge University Press.

Perry, S., and L. M. Rose. 1994. Begging and transfer of coati meat by white-faced capuchin monkeys, *Cebus capucinus. Primates* 35: 409–415.

Plummer, T., and C. B. Stanford. Submitted. Analysis of a prey bone assemblage made by wild chimpanzees in Gombe National Park, Tanzania. *Journal of Human Evolution.*

Remis, M. J. 1997. Ranging and grouping patterns of a western lowland gorilla group at Bai Houkou, Central African Republic. *American Journal of Primatology* 43: 111–133.

Rodman, P. S., and H. M. McHenry. 1980. Bioenergetics and the origin of hominid bipedalism. *American Journal of Physical Anthropology* 52: 103–106.

Rose, M. D. 1984. Food acquisition and the evolution of positional behavior: The case of bipedalism. In *Food Acquisition and Processing in Primates,* ed. D. J. Chivers, B. A. Wood, and A. Bilsborough, 509–524. New York: Plenum Press.

Rose, L. M. 1997. Vertebrate predation and food-sharing in *Cebus* and *Pan. International Journal of Primatology* 18: 727–765.

Sabater Pi, J., M. Bermejo, G. Ilera, and J. J. Vea. 1993. Behavior of bonobos (*Pan paniscus*) following their capture of monkeys in Zaïre. *International Journal of Primatology* 14: 797–804.

Sanday, P. 1981. *Female Power and Male Dominance: On the Origins of Sexual Inequality.* New York: Cambridge University Press.

Schaller, G. B. 1963. *The Mountain Gorilla.* Chicago: University of Chicago Press.

Schaller, G. B., and G. R. Lowther. 1969. The relevance of carnivore behavior to the study of early hominids. *Southwestern Journal of Anthropology* 25: 307–341.

Sept, J. M. 1992. Was there no place like home? A new perspective on early hominid sites from the mapping of chimpanzee nests. *Current Anthropology* 33: 187–207.

Sept, J. M. 1994. Beyond bones: Archaeological sites, early hominid subsistence, and the costs and benefits of exploiting wild plant foods in east African riverine landscapes. *Journal of Human Evolution* 27: 295–320.

Shipman, P. 1986. Scavenging or hunting in early hominics. *American Anthropologist* 88: 27–43.

Shipman, P., and A. Walker. 1989. The costs of becoming a predator. *Journal of Human Evolution* 18: 373–392.

Simmons, R. E., and L. Scheepers. 1996. Winning by a neck: Sexual selection in the evolution of giraffe. *American Naturalist* 148: 771–786.

Siskind, J. 1973. *To Hunt in the Morning*. New York: Oxford University Press.

Smith, E. A., and B. Winterhalder, eds. 1992. *Evolutionary Ecology and Human Behavior*. New York: Aldine.

Smuts, B. B. 1995. The evolutionary origins of patriarchy. *Human Nature* 6: 1–32.

Smuts, B. B., and R. W. Smuts. 1993. Male aggression and sexual coercion of females in nonhuman primates and other mammals: Evidence and theoretical implications. *Advances in the Study of Behavior* 22: 1–63.

Speth, J. D. 1987. Early hominid subsistence strategies in seasonal habitats. *Journal of Archaeological Science* 14: 13–29.

Speth, J. D. 1989. Early hominid hunting and scavenging: The role of meat as an energy source. *Journal of Human Evolution* 18: 329–343.

Speth, J. D. 1990. Seasonality, resource stress, and food sharing in so-called "egalitarian" foraging societies. *Journal of Anthropological Archaeology* 9: 148–188.

Speth, J. D. 1991. Protein selection and avoidance strategies of contemporary and ancestral foragers: Unresolved issues. *Philosophical Transactions of the Royal Society of London* 334: 265–270.

Speth, J. D., and D. D. Davis. 1976. Seasonal variability in early hominid predation. *Science* 192: 441–445.

Stanford, C. B. 1995. The influence of chimpanzee predation on group size and anti-predator behaviour in red colobus monkeys. *Animal Behaviour* 49: 577–587.

Stanford, C. B. 1996. The hunting ecology of wild chimpanzees: Implications for the behavioral ecology of Pliocene hominids. *American Anthropologist* 98: 96–113.

Stanford, C. B. 1998a. *Chimpanzee and Red Colobus: The Ecology of Predator and Prey*. Cambridge, Mass.: Harvard University Press.

Stanford, C. B. 1998b. The social behavior of chimpanzees and bonobos: Empirical evidence and shifting assumptions. *Current Anthropology* 39: 399–420.

Stanford, C. B., and J. S. Allen. 1991. Strategic storytelling: Current models of human behavioral evolution. *Current Anthropology* 32: 58–61.

Stanford, C. B., J. Wallis, H. Matama, and J. Goodall. 1994a. Patterns of predation by chimpanzees on red colobus monkeys in Gombe National Park, Tanzania, 1982–1991. *American Journal of Physical Anthropology* 94: 213–228.

Stanford, C. B., J. Wallis, E. Mpongo, and J. Goodall. 1994b. Hunting decisions in wild chimpanzees. *Behaviour* 131: 1–20.

Stephens, D. W., and J. R. Krebs. 1986. *Foraging Theory*. Princeton, N.J.: Princeton University Press.

Steudel, K. L. 1994. Locomotor energetics and hominid evolution. *Evolutionary Anthropology* 3: 42–48.

Steudel, K. 1996. Limb morphology, bipedal gait and the energetics of hominid locomotion. *American Journal of Physical Anthropology* 99: 345–355.

Stewart, K., and A. Harcourt. 1987. Gorillas: Variation in female relationships. In *Primate Societies*, ed. by B. B. Smuts, D. L. Cheney, R. M. Seyfarth, R. W. Wrangham, and T. T. Struhsaker, 146–164. Chicago: University of Chicago Press.

Stiner, M. C. 1991. An interspecific perspective on the emergence of the modern human predatory niche. In *Human Predators and Prey Mortality*, ed. M. C Stiner, 149–186. Boulder, Colo.: Westview Press.

Straus, L. G. 1987. Hunting in late Upper Paleolithic Western Europe. In *The Evolution of Human Hunting*, ed. M H. Nitecki and D. V. Nitecki, 147–175. New York: Plenum Press.

Strier, K. B. 1994. Myth of the typical primate. *Yearbook of Physical Anthropology* 37: 233–271.

Strum, S. C 1981. Processes and products of change: Baboon predatory behavior at Gilgil, Kenya. In *Omnivorous Primates*, ed. R.S.O. Harding and G. Teleki, 255–302. New York: Columbia University Press.

Strum, S. C. 1983. Baboon cues for eating meat. *Journal of Human Evolution* 12: 327–336.

Sugardjito, J., and N. Nuhuda. 1981. Meat-eating behavior in wild orangutans. *Primates* 22: 414–416

Takahata, Y., H. Ihobe, and G. Idani. 1996. Comparing copulations of chimpanzees and bonobos: Do females exhibit proceptivity or receptivity? In *Great Ape Societies*, ed. W. C. McGrew, L. F. Marchant, and T. Nishida, 146–155. Cambridge, U.K.: Cambridge University Press.

Tanner, N. M., and A. L. Zihlmann. 1976. Women in evolution, part 1: Innovation and selection in human origins. *Signs: Journal of Women, Culture, and Society* 1: 585–608.

Tappen, M. 1995. Savanna ecology and natural bone deposition: Implications for early hominid site formation, hunting, and scavenging. *Current Anthropology* 36: 223–270.

Taylor, C. R., and V. J. Rowntree. 1973. Running on two or four legs: Which consumes more energy? *Science* 179: 186–187.

Teleki, G. 1973. *The Predatory Behavior of Wild Chimpanzees.* Lewisburg, Pa.: Bucknell University Press.

Thompson-Handler, N., R. K. Malenky, and N. Badrian. 1984. Sexual behavior of *Pan paniscus* under natural conditions in the Lomako Forest, Equateur, Zaïre. In *The Pygmy Chimpanzee*, ed. R. L. Susman, 347–368. New York: Plenum Press.

Tiger, L. 1969. *Men in Groups.* New York: Vintage Books.

Tiger, L., and R. Fox. 1971. *The Imperial Animal.* New York: Holt, Rhinehart and Winston.

Tooby, J., and L. Cosmides. 1992. In *The Adapted Mind*, ed. J. Barkow, L. Cosmides, and J. Tooby, 19–136. Oxford: Oxford University Press.

Tooby, J., and I. DeVore. 1987. The reconstruction of hominid behavioral evolution through strategic modeling. In *The Evolution of Human Behavior: Primate Models*, ed. W. G. Kinzey, 183–238. Albany, N.Y.: State University of New York Press.

Toth, N. , K. Schick, and S. Savage-Rumbaugh. 1993. Pan the tool-maker: Investigations into the stone tool-making and tool-using capabilities of a bonobo (*Pan paniscus*). *Journal of Archaeological Science* 20: 81–91.

Trinkhaus, E. 1987. Bodies, brawn, brains and noses: Human ancestors and human predation. In *The Evolution of Human Hunting*, ed. M. H. Nitecki and D. V. Nitecki, 107–145. New York: Plenum Press.

Tutin, C.E.G. 1979. Mating patterns and reproductive strategies in a community of wild chimpanzees (*Pan trog-lodytes schweinfurtnii*). *Behavioral Ecology and Sociobiology* 6: 29–38.

Tutin, C.E.G. 1996. Ranging and social structure of lowland gorillas in the Lopé Reserve, Gabon. In *Great Ape Societies*, ed. W. C. McGrew, L. F. Marchant, and T. Nishida, 58–70. Cambridge, U.K.: Cambridge University Press.

Tuttle, R. H. 1974. Darwin's apes, dental apes, and the descent of man: Normal science in evolutionary anthropology. *Current Anthropology* 15: 389–398.

van Schaik, C. P., and J.A.R.A.M. van Hooff. 1983. On the ultimate causes of primate social systems. *Behaviour* 85: 91–117.

van Schaik, C. P., and J.A.R.A.M. van Hooff. 1996. Toward an understanding of the orangutan's social system. In *Great Ape Societies*, ed. W. C. McGrew, L. F. Marchant, and T. Nishida, 3–15. Cambridge. U.K.: Cambridge University Press.

Wallis, J. 1997. A survey of reproductive parameters in the free-ranging chimpanzees of Gombe National Park. *Journal of Reproduction and Fertility* 109: 297–307.

Washburn, S. L. 1959. Speculations on the interrelations of the history of tools and biological evolution. *Human Biology* 31: 21–31.

Washburn, S. L. 1967. In *Anthropology Today*, ed. A. L. Kroeber. Chicago: University of Chicago Press.

Washburn, S. L., and C. Lancaster. 1968. The evolution of hunting. In *Man the Hunter*, ed. R. B. Lee and I. DeVore, 293–303. Chicago: Aldine.

Watts, D. P. 1989. Infanticide in mountain gorillas: New cases and a reconsideration of the evidence. *Ethology* 81: 1–18.

Wheeler, P. E. 1984. The evolution of bipedality and loss of functional body hair in hominids. *Journal of Human Evolution* 13: 91–98.

White, F. J. 1988. Party composition and dynamics in *Pan paniscus*. *International Journal of Primatology* 9: 179–193.

Winterhalder, B. 1996. Social foraging and the behavioral ecology of intragroup resource transfers. *Evolutionary Anthropology* 5: 46–57.

Winterhalder, B. 1997. Gifts given, gifts taken: The behavioral ecology of nonmarket, intragroup exchange. *Journal of Archaeological Research* 5: 121–170.

Wood, K. D., and F. J. White. 1996. Female feeding priority without female dominance in wild pygmy chimpanzees. *American Journal of Physical Anthropology*, supp. 22: 247 (abstract).

Woodburn, J. 1968. An introduction to Hadza ecology. In *Man the Hunter*, ed. R. B. Lee and I. DeVore, 49–55. Chicago: Aldine.

Wrangham, R. W. 1980. An ecological model of female-bonded primate groups. *Behaviour* 75: 262–292.

Wrangham, R. W. 1987. The significance of African apes for reconstructing human social evolution. In *The Evolution of Human Behavior: Primate Models*, ed. W. G. Kinzey, 51–71. Albany: State University of New York Press.

Wrangham, R. W. 1993. The evolution of sexuality in chimpanzees and bonobos. *Human Nature* 4: 47–79.

Wrangham, R. W., and D. Peterson. 1996. *Demonic Males*. Boston: Houghton Mifflin.

Yamagiwa, J., T. Maruhashi, T. Yumoto, and N. Mwanza. 1996. Dietary and ranging overlap in sympatric gorillas and chimpanzees in Kahuzi-Biega National Park, Zaïre. In *Great Ape Societies*, ed. W. C. McGrew, L. F. Marchant, and T. Nishida, 82–98. Cambridge, U.K.: Cambridge University Press.

Yellen, J. E. 1991. Small mammals: !Kung San utilization and the production of faunal assemblages. *Journal of Anthropological Research* 10: 1–26.

Zihlmann, A. L., J. E. Cronin, D. L. Cramer, and V. M. Sarich. 1978. Pygmy chimpanzee as a possible prototype for the common ancestor of humans, chimpanzees and gorillas *Nature* 275: 744–746.

Index